T0264877

DRINKING WATER
DISINFECTION
TECHNIQUES

DRINKING WATER DISINFECTION TECHNIQUES

**Aniruddha Bhalchandra Pandit
and Jyoti Kishen Kumar**

CRC Press
Taylor & Francis Group
Boca Raton London New York

CRC Press is an imprint of the
Taylor & Francis Group, an **informa** business

CRC Press
Taylor & Francis Group
6000 Broken Sound Parkway NW, Suite 300
Boca Raton, FL 33487-2742

First issued in paperback 2017

© 2013 by Taylor & Francis Group, LLC
CRC Press is an imprint of Taylor & Francis Group, an Informa business

No claim to original U.S. Government works

ISBN-13: 978-1-4398-7740-1 (hbk)
ISBN-13: 978-1-138-07387-6 (pbk)

This book contains information obtained from authentic and highly regarded sources. Reasonable efforts have been made to publish reliable data and information, but the author and publisher cannot assume responsibility for the validity of all materials or the consequences of their use. The authors and publishers have attempted to trace the copyright holders of all material reproduced in this publication and apologize to copyright holders if permission to publish in this form has not been obtained. If any copyright material has not been acknowledged please write and let us know so we may rectify in any future reprint.

Except as permitted under U.S. Copyright Law, no part of this book may be reprinted, reproduced, transmitted, or utilized in any form by any electronic, mechanical, or other means, now known or hereafter invented, including photocopying, microfilming, and recording, or in any information storage or retrieval system, without written permission from the publishers.

For permission to photocopy or use material electronically from this work, please access www.copyright.com (http://www.copyright.com/) or contact the Copyright Clearance Center, Inc. (CCC), 222 Rosewood Drive, Danvers, MA 01923, 978-750-8400. CCC is a not-for-profit organization that provides licenses and registration for a variety of users. For organizations that have been granted a photocopy license by the CCC, a separate system of payment has been arranged.

Trademark Notice: Product or corporate names may be trademarks or registered trademarks, and are used only for identification and explanation without intent to infringe.

Library of Congress Cataloging-in-Publication Data

Jyoti, K. K. (Kishen Kumar)
    Drinking water disinfection techniques / A.B. Pandit, K.K. Jyoti.
        p. cm.
    Includes bibliographical references and index.
    ISBN 978-1-4398-7740-1 (hardback)
    1. Drinking water--Purification. I. Pandit, A. B. (Aniruddha Bhalchandra) II. Title.

TD430.J96 2012
628.1'62--dc23   2012035001

Visit the Taylor & Francis Web site at
http://www.taylorandfrancis.com

and the CRC Press Web site at
http://www.crcpress.com

To my father

Professor Bhalchandra R. Pandit (alias Nana)

Aniruddha Bhalchandra Pandit (alias Vishwesh)

To my beloved parents

Smt. Lakshmi and Shri. P.A. Gopalkrishnan (Amma and Appa)

Jyoti Kishen Kumar

Contents

Foreword

Water is essential for humans, cattle, agriculture, power generation, industry, and so forth, and with the global population approaching 8 billion by the end of 2030, it will become even more important. The cotton textile industry and pulp and paper industry have traditionally consumed large amounts of water without proper recovery or recycling. It appears that almost 70% of the freshwater supply globally goes to agriculture. Power generation is yet another large consumer of water. In India, with its large population, and similarly in China, there are additional problems of sanitation, and enteric fever is still common in India, primarily due to the poor quality of drinking water in many areas. Jyoti and Pandit have done a highly commendable job in bringing about this book, collating disparate subjects in an orderly way. Chemistry plays a pivotal role in dealing with water treatment, and the use of flocculants and coagulants, softening salts, ion exchange resins (IERs), and so forth, plays an important role. Just imagine how IERs can selectively remove unwanted ions at room temperature, and of course, water treatment for high-pressure boilers stands out as an example where the dissolved solids have to be brought down to below 5 ppb. The global water treatment industry is probably worth around $350 billion per annum.

The advent of membranes, particularly hollow fiber based, has made a major impact as brackish water and seawater can be converted at room temperature into potable water. It appears that even injectable-grade water can be made via membrane processes. These technologies, with all types of membrane processes, such as microfiltration, ultrafiltration, reverse osmosis, and now nanofiltration (NF), have implications in terms of large energy savings. The growth of NF is impressive, as all unwanted organic impurities, including antibiotics, agrochemicals, dyes, and so on, with a molecular weight of more than 250 can be so easily and effectively removed. Development of low-energy membranes will provide further impetus. Saudi Arabia consumes almost 50% of its water through desalination.

The use of municipal wastewater is gaining importance, and recycling, with proper treatment, is becoming a necessity. Europe provides an excellent background for this subject, for example, with Rhine water usage going through different countries with strict conditions.

The power industry also consumes a large amount of water, and there are many ways of conserving and recycling. Many chemical industries, such as BASF, Bayer, Dow, and others have demonstrated decreased water consumption in the past few years, even with escalating levels of production. For instance, Dow operates on three Rs; reduce through conservation, recycle, and reuse. Municipal water recycling is cheaper than desalination.

Although the title of this book refers to drinking water disinfection, it covers much wider areas. There is a reference to extensive use of biocides and how chlorine plays an important role and how the chemical industry has dealt with making Cl_2 via the Deacons process, from HCl to electrolysis of brine, now using membrane cells. Sodium hypochlorite is used in several ways, including via online electrolysis even of seawater. It is possible to make pure sodium hypochlorite from HClO, free from chlorite, in turn made by absorption of chlorine in hot caustic soda with efficient desorption of HClO. A variety of biocides are on the market, and the quest for new ones continues. Many unwanted products from the chemical industry, such as aluminium chloride, ferrous sulphate, and dilute sulphuric acid, have been converted into good coagulants like polyaluminium chloride, polyferricsulphate, and alum, respectively.

Pandit has done extraordinary work using cavitation and ultrasonication for a variety of purposes, and apart from publications in learned journals, has established its industrial applications, including disinfection. The chapter on cavitation is particularly praiseworthy.

I congratulate the authors in undertaking this daunting task. I believe that this book will find a place on the shelves of anyone interested in this area.

Padma Vibhushan Professor M.M. Sharma, FNA, FRS
Former Director and Emeritus Professor of Eminence
Institute of Chemical Technology
Mumbai, India

Preface

The basic idea of writing a book on water treatment for potable water has been in the back of my mind since the late 1990s, when I (J.K.K.) started working on my Ph.D. Around the same time, a big rainwater harvesting initiative was undertaken by the institute, due to the depletion of the groundwater level. However, the bore wells dug in the institute started giving brackish/saline water, which could not be used even for gardening.

Thus, the work of water harvesting and treatment started simultaneously to get water, initially of gardening quality and subsequently of a potable standard. The approach chosen for these two qualities of water was (1) desalination/softening for reducing the total dissolved solids (TDS) and (2) disinfection to reduce the microbial load. The authors successfully came up with strategies to get potable water by combining physical and chemical methods, and the economic comparison of these techniques was a significant driver for this book. Our initial apprehension about the acceptability of such a book was completely unfounded, once we received the comments from one of the reviews. That person stated, "Although the technology of water disinfection is well established, there is a need of updated and comprehensive information, clearly evidencing the advantages but mainly the disadvantages of each (treatment) choice." This review also essentially indicated the expected focus and the target audience for the book.

The focus of the book is to address the issue of the microbial disinfection of water to make it potable. However, disinfection and consumption are many times separated by days/weeks/months on a timescale, and hence, the technique chosen for disinfection should also have a lasting (long-term) effect. In a country like India, the water precipitated during the monsoon months (June to September) is stored in huge reservoirs and supplied after treatment for the rest of the year. Thus, the quality of source water over a period keeps changing due to its exposure to the atmosphere, contamination from other industrial activities in the vicinity of the reservoir, and so forth. As a result, the treatment plant needs to be continuously/periodically tuned to adjust to the quality of the source water. This is a significant challenge, as the treatment has to take into account not only the changing microbial load, but also other organic and inorganic pollutants and their effect on the existing microbial disinfection techniques, as well as the destruction of these pollutants so that they do not act as substrates for subsequent microbial growth to make water of potable (drinking) quality.

Although microorganisms always existed, their adverse effect as pathogens and water as a carrier was addressed only in the eighteenth and nineteenth centuries. Waterborne diseases continue to present challenges to public health officials and water suppliers. Though water purification, through boiling, filtration, and sedimentation, has been practiced since ancient times, its scale of operation and engineering analysis became important as urbanisation became significant. There emerged new concerns that conventional treatment methods may not be adequate to address the issue of emerging pathogens and viruses, and similar to the development of higher-generation antibiotics to destroy the mutated pathogens, the treatment strategies required continual modification/improvement. As new techniques of treatment are evolving, supported by accurate and very sensitive analytical techniques, disinfection by-products (DBPs) as a function of a new treatment scheme have become an important assessment parameter (in addition to economies) regarding the final choice of a treatment or combinations of treatments.

Thus, this book broadly addresses the issues related to water disinfection in the form of three generic areas: (1) disinfection techniques—chemical, physical, and hybrid (combination)—and their likely scale of operation efficacy; (2) DBP formation, as a function of water source and the type of treatment and the DBP's adverse effect; and (3) emerging/novel techniques, which are the future of this technology as per the authors' assessment.

These three broad areas discuss issues related to microbiology, chemistry, and engineering (limited to the principle of operation and cost of technology and the treatment) and offer a global perspective to the reader. The standard analytical tools and recommended methods to be used for all of these treatment schemes have also been discussed, along with appropriate references for further reading. The history and evolution of the treatment standard has been presented, along with the current and future challenges that need resolution.

The topics discussed are still continuously evolving and substantial work is being done and published. We have tried our best to note and discuss the recent work; however, some deficiencies can always be noted, which hopefully will be taken care of in due course.

Professor Aniruddha Bhalchandra Pandit

Dr. Jyoti Kishen Kumar

Acknowledgments

A book such as this, on a topic of such global importance, would not have been possible without the support of many people. We thank Professor M.M. Sharma for his foreword and constant encouragement. His periodic nudging has always been a source of inspiration throughout our academic careers. Professor J.B. Joshi has always been a friend, philosopher, and guide and has given correct tips and direction to our research efforts. The area of water treatment for future research was his suggestion, which has now culminated in the form of this book. We also thank the Taylor & Francis/CRC Press team for their support and for publishing this book in the best fashion.

Professor Aniruddha Bhalchandra Pandit acknowledges the University Grants Commission (government of India) and the Department of Science and Technology (government of India) for their awards and project funding to research this area. We also acknowledge our colleagues, Dr. B.D. Kulkarni, Professor V.G. Gaikar, and Dr. P.R. Gogate for interesting and stimulating discussions during the course of this book writing, and our students, namely, Shree Dipak Pinjari, Kiran Ramisetty, Yogesh Shinde, and Gaurav Dastane, for compiling and typing some of the written parts and comments to bring more clarity to the discussion.

Professor Pandit also acknowledges the love and support of his wife, Anala, daughter, Sphoorti, and in-laws, Aai and Anna, for never complaining about his absence from many family functions due to this commitment to the book.

Jyoti Kishen Kumar expresses her gratitude to Professor Aniruddha Bhalchandra Pandit for his professional advice, valuable contribution, and guidance to see her through the book. Professor S.S. Lele has been a friend, philosopher, and guide and always encouraged this endeavor. This book would not have been possible without the love and support of family and friends, especially my husband, Kumar, and in-laws for understanding my frequent short spells at the computer, and my dearest daughters, Ranjini and Rajashree, who wholeheartedly cooperated and understood in spite of all the time it took away from them.

1

Introduction

1.1 Current Scenario

Although three-quarters of the earth's surface is water, only 1% is available for direct use, including drinking, and this often requires treatment before it can be used safely. Water contains many kinds of microbes and organisms, which can cause disease. It is estimated that 80% of all sickness and disease in developing countries is caused by unsafe water and inadequate sanitation. Waterborne diseases continue to present challenges to public health officials and water suppliers. Prevention and control of waterborne diseases through water source protection and proper treatment techniques are critically important. Untreated or inadequately treated drinking water supplies, primarily from surface water sources, contain microorganisms that can cause outbreaks of waterborne diseases. There are new concerns about emerging pathogens, including protozoa like *Giardia*, *Cryptosporidium*, and certain viruses that might be found even in drinking water treated by conventional methods. Vulnerable populations such as the young, the elderly, and those with compromised immune systems remain at risk for significant illness and even death. If left untreated, drinking water supplies (primarily from surface water sources) will cause waterborne diseases such as cholera, typhoid, and dysentery. Over the last century, it has been observed that water use has increased at more than double the rate of population rise. It has been predicted by the United Nations that 1800 million people will be living in regions of absolute water scarcity and two-thirds of the world population could be under stress conditions by the year 2025. The water withdrawals by the developing countries would increase to 50%, and those by developed countries would be around 18% (Water Use, 2012).

Disinfection of drinking water and treatment of municipal wastewater provide critical public health protection. Disinfection destroys bacteria and viruses, helping to protect ecosystems and prevent the spread of waterborne disease. The most commonly used disinfectant for both drinking water and wastewater treatment is chlorine. Its effectiveness against a wide spectrum of disease-causing organisms, relatively low cost, and high reliability contribute to its popularity. Chlorine can be applied to water directly as a gas, or through the use of chlorinating chemicals. A number of alternative disinfectants, such as chlorine dioxide, chloramines, ozone, and ultraviolet radiation, are also used to varying degrees. Each disinfection technology has unique benefits, limitations, and costs. Individual water treatment system operators must weigh these trade-offs and choose disinfection methods based on local water quality conditions and the needs and resources of the communities they serve. Safe drinking water is of utmost importance globally. In fact, one of the Millennium Development Goals (MDGs) set by the United Nations is to reduce by half the proportion of people unable to reach or afford safe drinking water by 2015.

1.2 The Aquatic Environment

Water contains a variety of microorganisms, some of which are harmless, whereas some can cause fatal diseases in humans. Since most water contains nutrients to support these microorganisms, the microbial variety differs only according to the source of the water. Among the wind array of organisms are the algae, which derive the mineral nutrients from water and energy from sunlight. Bacteria that are present in water may be photolithic or chemolithic. They are also able to grow in environments with low nutrient concentrations. Once the autotrophs grow, a succession of heterotrophs emerges to decompose the organic material of the dead autotrophs and other organic matter. Lakes and rivers that are not polluted by sewage contain nutrients that are lower than in polluted waters, and the number of bacteria present in such water is also limited. These include soil saprophytes like *Micrococcus, Bacillus, Proteus,* and *Pseudomonas* species. Other types of microorganisms that can grow at zero or subzero temperatures are called psychrophilic bacteria, and these include *Aeromonas, Vibrio,* and *Clostridium* species. Apart from these bacteria, sulphur bacteria and many sulphate-reducing species are also found. These bacteria (prokaryotes) do not have a well-defined nucleus like the eukaryotes, but their nuclear material is present in the form of a mass called the *nucleoid*. The bacterial cells are covered by a bacterial membrane, which is of prime importance when attempting the disinfection of water, as it is the target of many disinfection agents. This means that the structure of the bacterial membrane is a parameter of major relevance in water disinfection.

The bacterial cells are bound by a surface membrane known as the cytoplasmic membrane. The cytoplasmic membrane is composed of lipids and proteins embedded within them. The lipid consists of both hydrophobic and hydrophilic regions and hence is amphoteric in nature. They are oriented in such a manner that the hydrophobic regions are within the membrane so that the water is excluded from the membrane, whereas the hydrophilic portion lies facing the aquatic environment on both sides. In a similar manner, the proteins that are embedded within the lipids also contain hydrophobic amino acid residues, which lie within the membrane, and the hydrophilic residues are exposed on both sides.

In a typical bacterium, the rigidity and strength of the cell wall can be attributed to peptidoglycans or mucopeptides. The peptidoglycans consist of alternating units of N-acetylglucosamine and N-acetylmuramic acid with β-1-4 linkages. The muramic acid is linked to short peptide, which varies in composition but always contains a minimum of three amino acids: alanine and glutamic acid, either diaminopilemic acid or the structurally related amino acid lysine. The structural rigidity of the peptidoglycan is due to the cross-linking of the polymers. The type and extent of the cross-linkage varies among different species. For example, the peptidoglycan of the gram-positive bacteria is more extensively cross-linked than that of the gram-negative bacteria. The cell walls of the gram-negative bacteria are thin (10 to 15 nm) and make up to 20% of the dry weight of the cell. The gram-positive cell walls are thicker (25 to 30 nm) and make up to 20 to 40% of the dry weight (Frobisher et al., 1974). The cell membrane of bacteria has many important functions to perform. It acts as an osmotic barrier and prevents the entry of large molecules into the cell, thereby maintaining a favourable osmotic pressure within the cell. Second, a number of active transport systems necessary for the survival of the cell are present in this membrane. It also acts as a site for many enzymatic reactions involved in the energy metabolism of the cell (Frobisher et al., 1974).

1.3 Health Aspects

Pollution from sewage is an indication of poor hygiene and leads to the introduction of a number of microorganisms in water that, when consumed as such, may cause diseases ranging from mild gastroenteritis to severe and potentially fatal dysentery and cholera. Table 1.1 lists the diseases caused by some microbes present in water.

One of the important groups of bacteria found in the intestinal tract is the family Enterobacteriaceae, the enteric bacteria. *Salmonella* and *Shigella* belong to this group. *Proteus* and *Klebsiella* also belong to this group and are opportunistic or occasional pathogens. Bacteria like *Escherichia* and *Enterobacter* are present normally in the human intestine, but they cause diseases only under very exceptional conditions. *Escherichia* is regarded as a genus with only one species in which there are several hundred different antigenic types. Clinical isolates of *Escherichia coli* may be conveniently grouped into three categories: opportunistic, enteropathogenic, and enterotoxin producing. The opportunistic type are generally harmless

TABLE 1.1

Microbes and Health Hazards

Serial No.	Type of Microorganism	Name of Causative Microbe	Disease Caused	Association of Disease to Water Quality
1	Bacteria	*Salmonella typhosa*	Typhoid	These bacterial diseases are very strongly associated with unsafe potable water. Also related to unsanitary sewage disposal and poor hygiene.
		Salmonella flexneri	Dysentery	
		Vibrio cholerae	Cholera	
		Escherichia coli	Gastroenteritis	
2	Parasites	*Entamoeba histolytica*	Amoebiasis	Strongly connected with poor domestic and personal hygienic conditions and unsanitary sewage disposal.
		Ascario lumbricoides	Ascariasis	
		Schistosoma mansoni	Schistosomiasis	
		Giardia lamblia	Giardiasis	
		Cryptosporidium parvum	Cryptosporidiosis	
3	Viral	Poliovirus	Muscular paralysis	Related to unsanitary conditions and poor personal and domestic hygiene, consumption of unsafe drinking water.
		Hepatitis virus	Infectious hepatitis	

in their normal habitat until they gain access to other sites and tissues and can then cause infections of the urinary tract, meningitis, and pulmonary, skin, and wound infections. Enteropathogenic *E. coli* are pathogenic within the intestinal tract, causing acute gastroenteritis in newborns and infants but rarely in adults except those with lowered resistance to the infections. These organisms are unable to invade the intestinal mucosa, but they release an enterotoxin that adsorbs to the epithelial cell membranes and stimulates adenyl cyclase activity (cyclic AMP), which in turn leads to increased electrolyte secretion in the gut. Enteric fever or gastroenteritis is also caused by *Salmonella*. Dysentery of bacterial etiology is distinguished from other types by the term *shigellosis*, which is caused by different species of *Shigella*. Typhoid fever is an acute infectious disease caused by *Salmonella typhi*. This occurs due to improper disposal of biological wastes, improper purification of water, and poor sanitation practices. Other diseases arising due to polluted water are cholera and hepatitis. Cholera is caused by *Vibrio cholerae*. This disease is contagious and is transmitted through water and food contaminated with the excreta from patients and convalescent carriers. Viral hepatitis, which affects the liver, is caused by two viruses designated as hepatitis type A and hepatitis type B. Diseases caused by type A are transmitted through contaminated food and water. Yet another disease transmitted by infected food and water is amoebiasis, caused by *Entamoeba histolytica*, which results in severe dysentery. These diseases can be controlled by proper sanitation. Thus, drinking water must be wholesome and palatable. Accordingly, it must be free of microorganisms and poisonous undesirable substances, on one hand, and must be attractive to the senses, on the other hand.

Water quality has been deteriorating progressively over the years due to the dumping of wastes from industry, agricultural activities involving complex chemicals, and use and reuse of water without treatment. As a result of this the water becomes polluted with various chemical, physical, and biological matter, and thus becomes unfit for human consumption. Efforts must be made to improve the water quality, and a balance between water supply and water demand should be maintained through efficient management. This can be achieved by controlling the pollution from industries and agriculture, on one hand, and by finding effective ways of water treatment, on the other hand. There still exists a major challenge in measuring the scope of waterborne diseases. A new indicator, namely, disability-adjusted life year (DALY), was introduced to quantify the burden of the waterborne disease in 1993 by the Harvard School of Public Health in collaboration with the World Health Organisation (WHO) and World Bank. DALY indicates the number of years lost due to premature death and years of life lived with disabilities related to waterborne diseases. One DALY is equivalent to 1 year of healthy life lost. Among millions of lives lost, the majority appears to be those of small children, caused mainly by water-related diarrhoea. However, not much progress has been made since 2000 in curbing DALYs (www.unwater.org).

1.4 History of Disinfection

Water is rightly called the *elixir of life*, and for centuries water quality has been associated with health. Our ancestors may not have known the actual cause of water-related disease, but they definitely assessed the quality of water by its appearance. Therefore, clear water was considered safe, whereas cloudy water, which we now call turbid water, was linked to ill health. As early as 4000 B.C., man has attempted to improve the taste and odour of water, and ancient Sanskrit and Greek writings report of filtration methods using sand and charcoal as a remedy. By 1500 B.C., Egyptians used alum as the first chemical to suspend particles and thereby reduce turbidity, followed by slow sand filtration in many parts of Europe by the 1800s. In the year 1855, a cholera epidemic resulted in several deaths in London. John Snow, an English doctor, discovered that the disease was caused due to consumption of water from a public well that was contaminated by sewage. At this point, the exact reason for the cholera outbreak could not be explained. By the 1880s, the "germ theory" of disease was firmly established by Louis Pasteur, who determined that microbes could transmit deadly diseases such as cholera through media like water. By the late nineteenth and early twentieth centuries, humans started recognising the importance of safe drinking water, and also understood the role of pathogens in causing waterborne infections. Disinfection became the need of the hour, and several scientific contributions were made in this arena. Chemical disinfectants such as chlorine and ozone were used in the United States and Europe. At this point authorities also realised the necessity for water regulations to ensure safe drinking water. The U.S. Public Services have been reported to have set the bacteriologic guidelines in 1914; they were subsequently revised and expanded in 1925, 1946, and 1962. This was followed by the Safe Drinking Water Act in 1974.

Since 1974, drinking water professionals have recognised the need to modify traditional chlorine disinfection processes in response to advances in knowledge, particularly about the disinfection by-product formation. Some disinfection by-products are known to produce long-term harmful health effects. In 1979, the U.S. Environmental Protection Agency (USEPA) adopted a trihalomethane (THM) regulation, limiting the allowable level of this carcinogenic disinfection by-product in drinking water. Water utilities and treatment have changed their operations to minimise THM formation without compromising public health protection. Changes in operations have included reducing the amount of chlorine used, shifting the point of chlorine application, changing the type of chlorine used, and removing the naturally occurring organic matter that reacts with chlorine to produce THMs. The USEPA proposed a new drinking water regulation for disinfectants and disinfection by-products in 1994. It is anticipated that further changes in practices will occur (recent standards for microbial limits are discussed as a separate section in this chapter), particularly in response to recently proposed reductions in allowable THM levels and new limits on other by-products. These changes in the operation and plant process design will continue as new developments in treatment technology, scientific knowledge, and regulation occur.

1.5 General Aspects of Disinfection

In addition to potential pathogens, raw water may contain contaminants that may interfere with the disinfection process or be undesirable in the finished product. These contaminants include inorganic and organic molecules, particulates, and other organisms, for example, invertebrates. Variations among these contaminants arise from differences in regional geochemistry and between ground- and surface water sources. Many inorganic and organic molecules that occur in raw water exert a demand, that is, a capacity to react with and consume the disinfectant. Therefore, higher-demand waters require a greater dose to achieve a specific concentration of the active species of disinfectant. This demand must be satisfied to ensure adequate biocidal treatment. Ferrous ions, nitrites, hydrogen sulphide, and various organic molecules exert a demand for oxidising disinfectants such as chlorine. The bulk of the nonparticulate organic material in raw water occurs as naturally derived humic substances, that is, humic, fulvic, and hymatomelanic acids, which contribute to the colour of the water. The structure of these molecules is not yet fully understood. However, they are known to be polymeric and to contain aromatic rings and carboxyl, phenolic, alcoholic, hydroxyl, and methoxyl functional groups. Humic substances, when consuming and reacting with the applied chlorine, produce chloroform ($CHCl_3$) and other THMs. Water, particularly surface waters, may also contain synthetic organic molecules whose demand for disinfectant will be determined by their structure. Ammonia and amines in raw water will react with chlorine to yield chloramines that do have some biocidal activity, unlike most products of these side reactions. If chlorination progresses to the breakpoint, that is, to a free chlorine residual, these chloramines will be oxidised, causing more added chlorine to be consumed before a specific free chlorine level is achieved. The nature of the demand reaction varies with the composition of the water and the type of the disinfectant. Removal of the demand substances leaves water with a lower requirement of chlorine for the disinfectant to achieve an equivalent degree of protection against the transmission of a waterborne disease.

Various treatments applied to the raw water to remove undesirable characteristics, like colour, taste, odour, or turbidity, may affect the ultimate microbiological quality of the finished water. Microorganisms may be physically removed or the disinfectant demand of the water altered. Presedimentation to remove suspended matter, coagulation with alum or other agents, and filtration reduce the organic material in the raw water, and thus the disinfectant demand. Removal of ferrous iron similarly reduces the demand for oxidising disinfectants, as will aeration, which eliminates hydrogen sulphide. Prechlorination to a free chlorine residual is practiced early in the treatment sequence as one method to alter taste- and odour-producing compounds, suppress the growth of the organisms in the treatment plant, remove iron and manganese, and reduce the interference of organic compounds in the coagulation process. The necessity for these treatments or others is determined by the characteristics of the raw water. The selection of one of the various methods to achieve a particular result will be based upon cost-effectiveness in the particular situation. When chlorination is used, the point of application in the treatment sequence of some of the above-mentioned procedures can affect the final THM content of the finished water. Reduction of precursors in the raw water by coagulation and settling prior to chlorination generally reduces final THM production. The available information on these variations is limited, and a universally applicable procedure cannot be recommended in view of the diverse treatments required for different raw waters.

Yet another important aspect is particulates and aggregates. To inactivate organisms in water, the active chemical species must be able to reach the reactive site within the organism or on its surface. Inactivation will not result if this cannot occur. Microorganisms in water may be adsorbed onto surfaces like clays, silt, and organic matter. Due to this, they may acquire physical protection. Such particles, with the adsorbed microorganisms, may aggregate to form clumps, affording additional protection. Organisms themselves may also aggregate or clump together so that the organisms that are within the interior of the clump are shielded from the disinfectant and are not inactivated. Organisms may also be physically embedded within particles of faecal material, within larger organisms such as nematodes, or, in the case of viruses, within human body cells that have been discharged in the faecal material. To disinfect water adequately, the water must be pretreated, to reduce the concentration of solid materials to an acceptable low level. The primary drinking water turbidity standard of nephelometric turbidity unit (NTU) is an attempt to ensure that the concentration of particulates is compatible with current disinfection techniques.

Residuals in water disinfection are of paramount importance. Water supplies are disinfected through the addition or dosage of a chemical or physical agent. With a chemical agent, such as a halogen, a given dosage should theoretically impart a predetermined concentration (residual) of the active agent in the water. From a practical point of view, most natural waters exert a demand" for the disinfectant, as discussed above, so that the residual in the water is less than the calculated amount based on the dosage. The decrease in the residual, which is caused by the demand, is rapid in most cases, but it may be prolonged until the residual eventually disappears. In addition, the chemical disinfecting agent may decompose spontaneously, thereby yielding substances having little or no disinfection ability and exerting no measurable residual. For example, ozone not only reacts with substances in water that exert a demand, but also decomposes rapidly. To achieve microbial inactivation with a chemical agent, a residual must be present for a specific time. Thus, the nature and level of the residual, together with the time of exposure, are important in achieving disinfection or microbial inactivation.

Because the nature of the dosage-residual relationship for natural waters cannot be reliably defined, the efficacy of the disinfection with a chemical agent must be based on a residual concentration and time-of-exposure relationship. Residual measurements are important and useful in controlling the disinfection process. By knowing the residual-time relationship that is required to inactivate pathogenic or infectious agents, one can adjust the dosage of the disinfecting agent to achieve the residual that is required for effective disinfection with a given contact time and the subsequent usage. Thus, the effectiveness of the disinfection process can be controlled and judged by monitoring or measuring the residual. Following disinfection of a water supply at a treatment plant, the water is distributed to the consumers. Residual is also important for continued protection of the water supply against subsequent contamination in the distribution system. Accidental or mechanical failures in the distribution system may result in the introduction of infectious agents into the water supply. In the presence of a residual, disinfection will continue and, as a result, offer continued protection to the users. Physical agents such as radiation may provide effective disinfection during application, but they do not impart any persistent residual to the water. The dosage of a chemical agent that is used to effect microbial inactivation should not be so great that it imparts a health hazard to the water consumer. From another point of view, the aesthetic quality of the finished water should not be impaired by the dosage of the chemical agent or the residual that is required for effective disinfection.

1.6 Water Disinfection Methods

Water that is free of disease-producing microorganisms and chemical substances deleterious to health is called potable. Water contaminated with either domestic or industrial wastes is called nonpotable or polluted (Pelczar et al., 1986). The principal operations employed in a water purification plant to produce water of a quality safe for human consumption are sedimentation, filtration, and disinfection. Disinfection of water can be achieved by various chemical and physical processes. Chemical processes comprise treatment of water with halogens, ozone, hypohalous salts, enzymes, and silver cations, while the physical processes comprise thermal treatment, application of ultrasound and electromagnetic radiation such as ultraviolet rays, x-rays, and γ- radiation, filtration through filters capable of retaining bacteria, and reverse osmosis. Adoption of these processes depends on the quality and quantity of water to be treated, location of the water supply, desired quality of treated water, safety, and economics. The method of choice for disinfecting water for human consumption depends on a variety of factors. These include its efficacy against waterborne pathogens, the accuracy with which the process can be monitored and controlled, its ability to produce a residual that provides an added measure of protection against possible post-treatment contamination resulting from faults in the distribution system, the aesthetic quality of the treated water, and the availability of technology for adoption of the method on a scale that is required for public water supplies. Economic factors will also play a part in the final decision. The propensity of various disinfection methods to produce by-products having effects on health (other than those relating to the control of infectious diseases) and the possibility of eliminating or avoiding these undesirable by-products are also important factors to be weighed when making the final decisions about over-all suitability of methods to disinfect drinking water. Typically, disinfection processes can be classified as chemical methods, which comprise treatment with chlorine (1–1.5 mg/l), chlorine dioxide, chloramines (0.5 mg/l), ozone (0.1–0.2 mg/l), and hydrogen peroxide (50–150 mg/l). Physical methods include heat treatment, ultraviolet light, and ultrasonication. A combination of the above processes for better efficiency and economics is termed hybrid methods. Each of these is dealt with in detail in a separate chapter.

1.7 Assessment of Disinfection

The goal of disinfection of public water supplies is the elimination of the pathogens that are responsible for waterborne diseases. The transmission of diseases such as typhoid and paratyphoid fevers, cholera, salmonellosis, and shigellosis can be controlled with treatments that substantially reduce the total number of viable microorganisms in the water. While the concentration of organisms in drinking water after effective disinfection may be exceedingly small, sterilisation (i.e., killing *all* the microbes present) is not attempted. Sterilisation is not only impractical but it cannot be maintained in the distribution system. Assessment of the reduction in microbial concentration that is sufficient to protect against the transmission of pathogens in water is discussed below.

Comparison of the biocidal efficacy of disinfectants is complicated by the need to control many variables, which was not realised in some early studies. Halogens in particular are significantly affected by the composition of the test solution and its pH, temperature, and halogen demand. For very low concentrations of halogen to be present over a testing period, halogen demand must be carefully eliminated. Different disinfectants may have different biocidal potentials. In earlier work, analytical difficulties may have precluded defining exactly the species present, but new techniques allow the species to be defined for most disinfectants. Investigators studying efficacy have usually adopted one of the two extremes. Some have conducted carefully designed laboratory experiments with controls for as many variables as possible. Some of these investigators have reduced the temperature to slow the inactivation reactions. Although these experiments yield good basic information and can be used to determine which variables are important, they often have little quantitative relationship to the field situations. The other extreme, a field study or reconstruction of field conditions, is difficult to control. Moreover, their results are often not reproducible. In addition to the variables noted above, prereaction of chemicals in the test system and the culture history of the organism being used may also affect the observed results. Despite these problems, there have been some attempts to standardise efficacy testing.

A major factor that influences the evaluation of the efficacy of a particular disinfectant is the test microorganism. There is a wide variation in susceptibility, not only among bacteria, viruses, and protozoa (cyst stage), but also among genera, species, and strains of the microorganism. It is impractical to obtain information on the inactivation by each disinfectant for each species and strain of pathogenic microorganism of importance in water. In addition, the condition and source of the test microorganism (e.g., the degree of aggregation and whether the organisms were naturally occurring or laboratory preparations), the presence of solids and particulates, and the presence of materials that react with and consume the disinfectant will affect the data interpretation. The overwhelming majority of the literature on water disinfection concerns the inactivation of model microorganisms rather than the pathogens. These disinfectant studies on model microorganisms have generally been nonpathogenic microorganisms that are as similar as possible to the pathogen and behave in a similar manner when exposed to the disinfectant. The disinfectant model systems are simpler, less fastidious, technically more workable systems that provide a way to obtain basic information concerning fundamental parameters and the biomolecular reactions. The information gained with the model systems can then be used to design key experiments in the more difficult systems. The model microorganisms used in the disinfection studies should be clearly distinguished from the indicator organism. The indicator should always be present when faecal material is present and absent in clean, uncontaminated water. It should die away in the natural aquatic environment and respond to treatment processes in a manner that is similar to that of the pathogens of interest. The indicator should be more numerous than the pathogens, and should be easy to isolate, identify, and enumerate. Only a restrictive application of the second criterion is necessary for a disinfection model. The response of the test microorganism to the disinfectant must be similar to that of the pathogen that it is intended to simulate. The disinfection model is not meant to function as an indicator microorganism. During the latter part of the nineteenth century, investigators recognised the presence of a group of bacteria that occurred in large numbers in faeces and wastewater. The most significant member of this group (currently called the coliform group) is *Escherichia coli*. Since the late nineteenth century, this coliform group has served as an indicator of the degree of faecal contamination of water, and *E. coli* has been used routinely as a disinfection model for enteric pathogens. The bacterial viruses of *E. coli* have received increased attention as possible disinfection models and indicators of enteric viruses in water and wastewater. Sommer et al. (2001) studied the inactivation of two bacteriophages, PHI X 174 and MS2, by UV and gamma radiation, in order to evaluate their potential as viral indicators for water disinfection by irradiation. Other researchers using chemical methods also carried out similar studies for, for example, iodine disinfection studies of a model bacteriophage, MS2 (Brion and Silverstein, 1999). Water has been documented as a vector of viral disease, but the agents primarily involved (hepatitis and gastroenteritis viruses) are difficult to detect in water. For practical purposes, such as monitoring of established treatment processes or the formulation of quality standards for water, a model organism would be desirable. A study conducted by an IAWPRC study group has proposed the use of bacteriophages as a model organism. Several procedures are available for detecting bacteriophages in water. Certain groups of bacteriophages, in particular the F-specific RNA bacteriophages, are relatively resistant to the conditions applied in the water treatment, and their resistance parallels that of some important groups of human viruses. It has been concluded that the presence of F-specific RNA bacteriophages is an index of sewage pollution, and consequently of the possible presence of human viruses. Several authors have shown a relation between densities of somatic coliphages and faecal coliforms (IAWPRC Study Group on Health-Related Water Microbiology, 1991). There is no generally accepted disinfection model for protozoan cysts. In disinfection studies for protozoan diseases, investigators have used the pathogen or its cysts. Work with such systems is, however, generally difficult.

The resistance or sensitivity to disinfectants of some bacteria (e.g., *E. coli*) in the laboratory may bear very little resemblance to their responses in nature. This is true in spite of the fact that standardised procedures govern the conditions under which cells are grown, harvested, washed, and so forth, when they are used as inocula. Examples of such differences include gram-negative bacteria, gram-positive bacteria, spores, and protozoa. Presumably, the mechanisms creating this phenomenon among these groups vary widely. A special situation is found with bacteria present in biofilms, which can be considered as being an intrinsic resistance mechanism resulting from physiological (phenotypic) adaptation of cells. Acquired resistance to biocides may arise by cellular mutation or by the acquisition of genetic elements (Russell, 1999). In halogen-disinfected waters, naturally occurring bacteria can be from one to two orders of magnitude more resistant to the disinfectant than cells of the same organism that had been subcultured on conventional laboratory culture media. Since standard disinfectant testing necessarily employs subcultured and washed bacterial cells, a false sense of confidence may be created if these data are used as an absolute criterion for the dilution of a disinfectant. These results could explain the frequent discrepancies between tests that are performed under laboratory conditions and those that are performed under field conditions. If bacteria could be used in their naturally occurring state, one might explore the possibility of bridging the gaps between laboratory and field conditions by using this experimental system. The ability of some gram-negative bacteria to grow in water makes it possible to produce and control large numbers of cells for such studies. More difficult to answer is the basic question of why naturally occurring cells of gram-negative water bacteria become more sensitive to disinfectants when grown in a minimal medium than the same strain when grown in water. One would expect the reverse to occur. The extreme resistance or differing resistances of naturally occurring bacteria can be attributed only to "environmental" factors and, perhaps, to the different compositions of cell walls and membranes. However, there have been no data to substantiate this hypothesis. Despite the questions that have been raised by differences in the behaviour of organisms under both laboratory and field conditions, valuable comparative information can be obtained from studies of disinfectants that are conducted in similar laboratory systems.

1.8 Quantitative Measurement of Bacterial Growth

The term *growth* as applied in the field of microbiology refers to the magnitude of the total population. Growth in this sense can be determined by various techniques based on cell count, which involves direct use of microscopy or an electron particle counter, or indirectly by a colony count performed by a pour or spread plate technique. Cell mass can also give an indication of growth. This is done directly by weighing or by measurement of cell nitrogen, or indirectly by turbidity measurements. Cell activity is yet another useful tool that is generally performed indirectly by relating the degree of biochemical activity to the size of the population. Each technique has its advantages and limitations, and no one method can be universally recommended. The colony count is the most widely used method in the general microbiological work. It should be emphasised that the colony count is theoretically the only technique that reflects the viable population. Moreover, discrepancies are likely to occur in the results of bacterial growth when measured by two different methods. For example, a microscopic count of a culture in the stationary phase would include all cells, viable and nonviable, whereas the colony count would reveal only the viable population (Pelczar et al., 1986).

The heterotropic plate count is a technique of colony count used to determine the general bacterial count of water. This method determines the number of cells that will multiply under certain defined conditions. A measured amount of the bacterial suspension is introduced into a petri dish, after which the agar medium (maintained in the liquid form at 45°C) is added and the two thoroughly mixed by rotating the plate. The plate count agar is usually the recommended medium for the standard bacterial plate count. However, other media such as Reasoner's 2A (R2A) media have been developed for the heterotropic plate count analysis of potable water samples (Reasoner and Geldreich, 1985). When the medium solidifies the organisms are trapped in the gel. Each organism grows, reproducing itself until a visible mass of organisms—a colony—develops; that is, one organism gives rise to one colony. Hence, a colony count performed on the plate reveals the visible microbial population of the inocula. The original sample is usually diluted so that the number of colonies developing on the plate will lie between 30 and 300. With this the count can be accurate and the possibility of interference from the growth of one organism with that of the other can be minimised. Colonies are usually counted by illuminating from below so that they are easily visible, and a large magnifying lens is often used. Various electronic techniques like image analysis have been developed for the counting of the colonies. This technique has some limitations, such as the ability of bacteria being tested to grow in the media being used under the test conditions. Also, cells that aggregate together may give rise to colonies that are lower than the individual cells, leading to false interpretation of the data. However, ease of performance and high sensitivity are the main merits, which highlight the use of the heterotropic plate count in water analysis to date.

The heterotropic plate count only gives information on the overall bacterial population. Although it is possible to detect many pathogens present in the water, the methods of enumeration are time-consuming and very complex. Moreover, pathogens are likely to gain entrance into water sporadically, and they do not survive for long periods of time; consequently, they could be missed out from a sample submitted to the laboratory. It is known that the pathogens that gain entrance into water bodies are via intestinal discharges of humans and animals. Certain bacterial species, particularly *E. coli* and related organisms designated as coliforms, faecal streptococci, and *Clostridium perfringens*, are normal inhabitants of the intestines of humans and other animals and are normally present in the faeces. Thus the presence of any of these species in water is evidence of faecal pollution of human or animal origin. If these organisms are present in water, then there is a possibility that intestinal pathogens can also gain entrance into humans through the consumption of water since they too occur in the faeces. Since the laboratory examinations of pathogens are troublesome, techniques have been developed for the demonstration of bacterial species of known excretal origin, particularly organisms of the coliform group (Pelczar et al., 1986).

These bacteria are called the indicator bacteria. An accepted indicator of the quality of drinking water is the coliform bacteria. The coliform group comprises all aerobic and facultative anaerobic, gram-negative, non-spore-forming, rod-shaped bacteria that ferment lactose with gas formation within 48 h at 35°C. Another indicator of the water quality is the faecal coliforms, which are a part of the coliform group. Faecal coliforms originate from the intestines of warm-blooded animals. They have properties similar to those of the coliform group, except that they grow at elevated temperatures of 44 to 44.5°C. The normal habitat of faecal streptococci is the intestines of humans and animals, and thus these organisms along with *E. coli* are also indicators of faecal pollution in water. Thus, its presence can provide valuable data on the bacteriological quality of water. In combination with the faecal coliform data, information on faecal streptococci may provide more specific data about pollution sources because certain faecal streptococci are host specific; that is, some of them may only infect animals, while some may be found only in humans. Faecal coliform/faecal streptococci ratios may provide information on the possible sources of pollution. A ratio greater than 4.1 is considered indicative of pollution derived from domestic wastes composed of human excrement, whereas ratios less than 0.7 suggest pollution due to nonhuman sources. Ratios between 0.7 and 4.4 usually indicate wastes of mixed human and animal sources (American Public Health Association, 1985). There are several analytical methods for the detection of the indicator microorganisms. Some of them are discussed in Table 1.2.

Although there is a plethora of analytical techniques to choose from, the drawbacks have to be understood well before selecting a particular technique. The heterotrophic plate count technique does not provide a good indication of the general load of the aerobic and the facultative anaerobic heterotrophic bacteria that treated water may carry because the conditions employed in the analytical technique may select for bacteria that may comprise only a small percentage of the bacterial population actually present. The remaining population is missed either because the microorganisms do not grow at all under the given conditions, or they grow very slowly and the colonies fail to reach a detectable size in the 48 h incubation period (Reasoner and Geldreich, 1985). Analytical problems could also be due to the viable but nonculturable bacteria. The bacterial population could undergo a decrease in plate count but remain viable when analysed by the direct viable count procedure. Bacteria that are known to exhibit this phenomena, that is, viable but nonculturable state, are of the genera *Vibrio, Escherichia, Salmonella, Aeromonas, Legionella, Campylobacter,* and *Shigella* (Byrd et al., 1991). Another problem is that of injured bacteria. Injury is the sublethal physiological consequence of exposure to stresses, which cause a loss in the ability of the microorganism to grow under normal conditions that are satisfactory for uninjured cells (LeChevallier and McFeters, 1985a). Bacteria become stressed in the aquatic environment as a result of low nutrient conditions. In water treatment processes and drinking water systems bacteria become stressed as a result of chemical and physical factors (McFeters et al., 1986). Such microbes will not grow under the normal test conditions, and hence this will lead to a false interpretation of the results. Interactions between the heterotrophic plate count bacteria and the coliform organisms can result in an injury to the coliform population. This can result in reduced coliform densities during the measurements (LeChevallier and McFeters, 1985b).

TABLE 1.2

Techniques Routinely Used for Assessing the Microbial Quality of Potable Water

Serial No.	Microbial Analytical Method	Methodology	Salient Points
1	Plate count (pour or spread plate)	Serially diluted sample is mixed with recommended media (generally the HPC agar) and incubated for 24 to 48 h.	Colonies counted are expressed as CFU/ml. It is a viable count. Recommended method by water authorities.
2	Most probable number (MPN) method	Consists of three steps: A presumptive test that is positive if gas formation occurs in series of culture media containing tubes inoculated with water sample, a confirmed test where the positive tubes are further grown in selective media to see gas formation, and finally, the completed test, where the positive samples are further streaked on special media plates.	The MPN of bacteria present can then be estimated using specially devised statistical tables, and the results are recorded with 95% confidence limits.
3	Membrane filter count	Use of molecular or membrane filters of 0.45 µm pore size. Filtration through the membrane retains most of the bacteria. Membrane with trapped microbes when placed on suitable media gives rise to colonies after incubation.	Variation on the plate count technique. Wide acceptance because the procedure is simple, rapid, and precise. Gives definite results. Factors like turbidity and membrane filter type may affect its sensitivity.
4	Clark's presence-absence test	Uses a single fermentation bottle containing the medium and a sample volume of either 50 or 100 ml. The bottles are incubated at 35 ± 0.5°C for up to 5 days and examined daily for growth (turbidity), acid, and gas production (Pipes et al., 1986).	Compared to the MPN method, which involves the use of multiple fermentation tubes, Clark's method is less cumbersome and can be easily applied to all types of water.

TABLE 1.2

Techniques Routinely Used for Assessing the Microbial Quality of Potable Water

Serial No.	Microbial Analytical Method	Methodology	Salient Points
5	The Colilert and Coliquik technique	Use of two active substrates, o-nitrophenyl-β-D-galactopyranoside (ONPG) and 4-methylumbelliferyl-β-D-glucuronide (MUG), to simultaneously detect total coliforms and *E. coli*. Total coliforms produce an enzyme β-galactosidase, which hydrolyses ONPG and releases o-nitrophenol, which produces a yellow colour. *E. coli* produces an enzyme β-glucuronidase, which hydrolyses MUG to form a fluorescent compound.	A comparative study of this method with the traditional membrane filter procedure indicated that both the techniques were acceptable for coliform detection (Olson et al., 1991).

1.9 History of Water Regulations

Water treatment processes have shown a significant improvement over the past century in an effort to minimise waterborne disease and transmission. In the United States, techniques have been developed in response to local conditions with other various objectives besides disinfection, including colour reduction, turbidity removal, softening, taste and odour control, and corrosion control. By the early twentieth century, the U.S. Public Health Service (USPHS) established bacteriological standards for drinking water. By the mid-twentieth century, the USPHS water quality standards were revised to include various chemical constituents as well as bacteriological indicators. This was significant since it extended the concern over water quality from waterborne diseases to toxicological (and eventually carcinogenic) effects from long-term ingestion. By the 1970s, with the passage of the Safe Drinking Water Act (1974), the U.S. Environmental Protection Agency (USEPA) was mandated to identify substances present in the drinking water that had adverse effects on the public health affects. As a result, national interim primary drinking water regulations were established by USEPA in 1977. In 1979, a group of chlorination by-products known as THMs were also

regulated. Over the next several years, continued health effects research aug-
mented the toxicological and carcinogenic database for various compounds.
In 1986, with the passage of the amendments to the Safe Drinking Water Act,
USEPA was mandated specifically to regulate microbiological constituents,
inorganic and organic compounds, and radioactivity to better safeguard
public health. These new regulations dramatically increased the monitoring requirements for public water systems. The latest amendments to the
Safe Drinking Water Act occurred in 1996. The criteria for the selection and
regulation of contaminants were a key amended item. Regulated contaminants either need to have adverse health effects or are present at levels sufficiently high to warrant public concern. Development of regulatory levels is
to be based on risk assessment, cost-benefit analysis, and minimising overall
risk. The law also takes into consideration the problems of small systems.
The best treatment technologies for smaller systems will be identified with
affordability as a major criterion. Small systems are also afforded more flex-
ibility with respect to monitoring and can be granted variances if the system
cannot afford to comply with the guidelines. Other specific areas addressed
include disinfection by-products, filtration, groundwater disinfection, arse-
nic, sulphate, radon, and source water protection (www.epa.gov).

1.10 Microbiological Standards

According to the USPHS, the raw water supply containing coliforms not in
excess of 5000/100 ml can, with modern water treatment processes, produce
potable water meeting the bacterial standards. Drinking water thus produced should not contain more than 1 coliform/100 ml. According to the
EEC Guidelines (1975), the maximum permissible limit for drinking water is
1000 count/100 ml, and for bathing water it is 10,000 coliforms/100 ml (Rump
and Krist, 1988). For outdoor bathing beaches the levels vary between 50 and
3000 coliforms/100 ml in different states, and for swimming the limit is 200
faecal coliforms/100 ml (Gehm and Bergman, 1976). As per the 2011 edition
of the *Drinking Water Standards and Health Advisories*, water disinfection systems that remove viruses, *Cryptosporidium*, and *Giardia lamblia* must achieve
99.9% inactivation. There are no limits mentioned specifically for *Legionella*
since it is believed to be controlled if most of the viruses and *Cryptosporidium*
and *Giardia lamblia* are successfully inactivated. The heterotropic plate count
(HPC) should not be more than 500 bacterial colonies/ml. The maximum
contaminant level (MCL) goal of total coliforms should be zero per 100 ml,
and the maximum contaminant level should not be more than 5% of the
samples being total coliform positive in a month. Every sample that has total
coliforms must be analysed for feacal coliforms, and there should be no feacal coliforms (http://www.epa.gov).

Questions

1. Why is water an important global issue today and what is the need for disinfecting water?
2. Describe the microorganisms routinely present in aquatic environments.
3. Explain the relevance of microbes in potable water with respect to health hazards.
4. What are the broad categories of water disinfection techniques?
5. Explain the various methods to access disinfection of water. Briefly describe each protocol.
6. Briefly explain the history of water regulations.
7. What are microbial standards?

References

American Public Health Association. (1985). *Standard methods for the examination of water and wastewater*, 16th ed. Washington, DC: APHA.

Brion, G.M., and Silverstein, J. (1999). Iodine disinfection of a model bacteriophage, MS2, demonstrating apparent rebound. *Water Res.* 33(1):169–179.

Byrd, J.J., Xu, H.-S., and Colwell, R.R. (1991). Viable but nonculturable bacteria in drinking water. *Appl. Environ. Microbiol.* 57(3):875–878.

Drinking water standards and health advisories. (2011). http://water.epa.gov/action/advisories/drinking/upload/dwstandards2011.pdf (accessed April 20, 2012).

Frobisher, M., Hinsdill, R.D., Crabtree, K.T., and Goodheart, C.R. (1974). *Fundamentals of microbiology*, 9th ed. Philadelphia: W.B. Saunders Company.

Gehm, W.H., and Bergman, J.I. (1976). *Handbook of water resources and pollution control.* New York: Van Nostrand Reinhold Company, p. 147.

IAWPRC Study Group on Health-Related Water Microbiology. (1991). Bacteriophages as model viruses in water quality control. *Water Res.* 25(5):529–545.

LeChevallier, M.W., and McFeters, G.A. (1985a). Enumerating injured coliforms in drinking water. *J. AWWA*, June, pp. 81–87.

LeChevallier, M.W., and McFeters, G.A. (1985b). Interactions between heterotropic plate count bacteria and coliform organisms. *Appl. Environ. Microbiol.* 49(5):1338–1341.

McFeters, G.A., Kippin, J.S., and LeChevallier, M.W. (1986). Injured coliforms in drinking water. *Appl. Environ. Microbiol.* 51(1):1–5.

Olson, B.H., Clark, D.L., Miller, B.B., Stewart, M.H., and Wolfe, R.L. (1991). Total coliform detection in drinking water: Comparison of membrane filtration with Colilert and Coliquik. *Appl. Environ. Microbiol.* 57(5):1535–1539.

Pelczar, M.J., Chan, E.C.S., and Kreig, N.R. (1986). *Microbiology*, 5th ed. Singapore: McGraw-Hill Book Co.

Pipes, W.O., Minnigh, H.A., Moyer, B., and Troy, M.A. (1986). Comparison of Clark's presence–absence test and the membrane filter method for coliform detection in potable water samples. *Appl. Environ. Microbiol.* 52(3):439–443.

Reasoner, D.J., and Geldreich, E.E. (1985). A new medium for the enumeration and subculture of bacteria from potable water. *Appl. Environ. Microbiol.* 49(1):1–7.

Rump, H.H., and Krist, H. (1988). *Laboratory manual for the examination of water, waste water and soil.* Weinheim: Verlag Chemie, pp. 161–164.

Russell, A.D. (1999). Bacterial resistance to disinfectants: Present knowledge and future problems. *J. Hosp. Infect.* 43(Suppl. 1):S57–S68.

Sommer, R., Pribil, W., Appelt, S., Gehringer, P., Eschweiler, H., Leth, H., Cabaj, A., and Haider, T. (2001). Inactivation of bacteriophages in water by means of non-ionizing (uv-253.7 nm) and ionizing (gamma) radiation: A comparative approach. *Water Res.* 35(13):3109–3116.

Water Use. (2012). http://www.unwater.org/statistics_use.html (accessed May 3, 2012).

2

Chemical Disinfection

2.1 Introduction

Chemical disinfection is used to destroy or control the growth microorganisms present in water that would otherwise cause fouling, corrosion of equipment, or lead to diseases from microbial activity (Hooper in Lorch, 1987). Disinfection is not sterilisation; that is, not all microorganisms are destroyed. However, the number of microbes left after disinfection should allow safe human consumption of the disinfected material. Hence, stringent guidelines laid down by water authorities have to be followed. This also implies that the guidelines for water disinfection should be chalked out taking into consideration the assessment of various physical, chemical, and bacteriological parameters. Generally, every country as well as individual state has regulations to ensure good drinking water quality. These rules and guidelines are reviewed and updated from time to time considering the changing global environmental scenario, including the emergence of new pathogens and pollutants as well as the sources of water. Yet another important element of water disinfection guidelines is to ensure safe residuals of the chemical used for disinfection given the fact that disinfection by-products are very harmful to humankind. This implies that drinking water that is disinfected per norms need not be completely devoid of deleterious by-products. Therefore, the use of chemical disinfection must be undertaken with care, as various limitations exist whereby products in water are sometimes unevenly distributed. In order to understand this, it is necessary to know some fundamentals about the theory of chemical disinfection.

2.2 Theory of Chemical Disinfection

Chemical disinfectants are nothing but antimicrobial agents that kill or inhibit the growth of microorganisms. Two groups of antimicrobial agents exist: oxidising and nonoxidising type agents. Oxidising agents include chlorine, chloramines, chlorine dioxide, chloride, bromide, bromines, ozone, and hydrogen peroxide. Nonoxidising agents comprise formaldehyde, iso-thiazalones, isocyanates, quaternary ammonium compounds, and chlorinated phenols (Hooper in Lorch, 1987). In general, antimicrobial agents kill microbes by acting either on the cell wall of the microbes or with the enzyme systems supporting the microbes' metabolic activity, or by both mechanisms. Needless to say, the exact mode of action will depend upon the type and concentration of disinfectant and the type of microbe being targeted. In addition, environmental factors will also play an important role. These aspects of disinfection will be dealt with in detail in the following sections. A successful water disinfection method depends on the efficiency of the disinfectant, and the relative efficiency of various disinfecting chemical compounds in turn depends on the rate of diffusion of the active chemical constituent through the cell wall. The critical factors that affect the disinfection efficiency are as follows (White, 1972):

- Nature of the disinfectant
- Concentration of the disinfectant
- Length of contact time between the disinfectant and the microbe
- Temperature
- Type and concentration of organisms
- pH and ionic strength

Disinfection by chemical agents has been researched extensively and its theory is explained easily with the help of mathematical models. Exposure of microorganisms to abnormal conditions (physical or chemical) will kill or arrest their growth and render them inviable. Death of a microorganism is defined as the irreversible loss of its ability to reproduce. When a microbial population is exposed to a lethal agent, the kinetics of death is usually exponential. This indicates that all members of the population are of similar sensitivity. The Chick–Watson model (Haas and Karra, 1984) is the most widely used tool to determine microbial inactivation by disinfectants. It is stated below:

$$\mathrm{Ln}\,(N/N_o) = -kC^n t$$

where:
N/N_o = Ratio of number of organisms at time t to time zero
C = Concentration of disinfectant (which must be constant, nondecaying) (mg)
n = Empirical constant or exponent
k = Death rate (CFU killed/min)
N_o = Initial number of organisms (CFU/ml)
N = Surviving number of organisms at a contact time, t (CFU/ml)
t = Duration of contact (min)

If the logarithm of the number of survivors is plotted as a function of time of exposure, a straight line should be obtained where the negative slope defines the death rate, K. The death rate alone only indicates the surviving fraction of the initial population over the treatment period. The underlying hypothesis of disinfection is (Langlais et al., 1991):

- The disinfectant concentration, C is a constant, nondecaying/decomposing during the reaction time, t.
- The microorganisms must be a single strain.
- The killing action must be of a single-hit and single-site type (one-time interaction).

Thus, disinfection kinetic models are the basis for assessing the disinfection performance of a disinfectant and the design of contactor systems (Trussell and Chao, 1977). Over the years, a number of kinetic models have been proposed for the formulation of disinfection design criteria mostly for chlorination. Model adequacy is dependent upon the robustness of the underlying inactivation rate law if the model accounts for the disappearance of the chemical disinfectant during the contact time. Several studies have been undertaken in this area, especially with regard to the dosage of disinfectant, contact time, prefiltration, and mixing intensity. It is very interesting to note that prefiltration of water samples with the membrane of 2.5 μm in pore size improved slightly disinfection efficiency. Mixing intensity did improve the disinfection efficacy (Lee and Nam, 2002). Having understood the importance of mathematical tools in water disinfection by chemicals, the following sections describe the remarkable chemical methods used for water treatment, beginning with the most conventional and established process of chlorination.

2.3 Chlorination

Chlorination is the most common water disinfection method that has been used for several decades globally. It is primarily used to destroy the pathogenic microorganisms in water. A secondary benefit of chlorination is odour removal to improve the quality of drinking water. According to the World Health Organisation (WHO), the adoption of drinking water chlorination has been one of the most significant advances in public health protection. Chlorine-based disinfectants are the only disinfectants that provide lasting residual protection to protect the water from waterborne disease throughout the distribution network, from the treatment plant to the consumer's tap. Alternatives to chlorination for primary disinfection, such as ozone or ultraviolet light, cannot provide this residual protection. As chlorination was adopted, human death rates due to cholera and hepatitis A also declined dramatically. Worldwide, significant strides in public health and the quality of life have been directly linked to the adoption of the chlorination of drinking water. Given below is a brief description of the history of chlorination.

The first milestone that leads to the development of chlorine as a disinfectant of choice is the discovery that microorganisms cause diseases. This was demonstrated by the scientific community in the late 1880s. This led to the realisation that antimicrobial compounds were required to prevent health hazards in humans. Since water is the basic necessity of man, its purity was imperative. This carved the way to the first application of chlorine disinfection to water facilities in England in the late 1890s, followed by the drafting of the first drinking water bacterial standard by the United States in the year 1915. By the end of 1918, several cities in the United States started employing chlorine disinfection. With emerging pathogens and the ever-increasing industrial pollution the standard had to be made stringent in the year 1925. By the end of the 1960s, a plethora of chlorine-based municipal water treating systems started operating throughout the United States. The U.S. Clean Water Act for restoring and maintaining the surface water was passed in 1972, followed by the U.S. Safe Drinking Water Act in 1974 and its amendments in 1986 and 1996.

Advantages of using chlorine are manifold. It is a very potent germicide as it reduces the levels of many disease-causing microbes in water to safe concentration levels, which adheres to the water standards. Chlorine-based disinfectants reduce many disagreeable tastes and odours. Chlorine oxidises many naturally occurring substances, such as foul-smelling algae secretions, sulphides, and odours from decaying vegetation. It also helps to remove iron and manganese from raw water by precipitation. One of the major advantages is the elimination of slime bacteria, moulds, and algae that commonly grow in water supply reservoirs, on the walls of the mains and storage tanks. Thus, chlorination offers very good biological growth control in distribution systems. The U.S. Environmental Protection Agency (USEPA) requires a residual level of disinfection of water in pipelines to prevent microbial regrowth and help protect treated water throughout the water distribution system. USEPA's maximum residual disinfection levels (MRDLs) are 4 mg/l for chlorine, 4 mg/l for chloramines, and 0.8 mg/l for chlorine dioxide (Drinking Water Standards, 2004). Although chlorine levels are usually significantly lower in tap water, USEPA believes that levels as high as the MRDLs pose no risk of adverse health effects, allowing for an adequate margin of safety (Drinking Water Standards, 2004).

2.3.1 Breakpoint Chlorination

Hypochlorous acid and other chlorine compounds having disinfecting ability by virtue of their being oxidising agents oxidise sulphites (SO_3^{2-}), sulphides (S$^-$), and ferrous (Fe^{2+}) or manganous (Mn^{2+}) ions. In effect, the disinfecting species are reduced, and the products have no disinfecting activity. All of the interfering compounds that destroy the disinfecting ability of the added chlorine exert a "chlorine demand," which may be defined as the difference between the amount of chlorine added and the quantity of free or combined available chlorine residual, measured in the water at the end of a specified contact period. When chlorine is added to the water with no chlorine demand, a linear relationship is established between the chlorine dosage and the free chlorine residual. However, when increasing amounts of chlorine are added to the water containing reducing agents and ammonia, the so-called breakpoint phenomenon occurs. The breakpoint is that dosage of chlorine that produces the first detectable amount of free available chlorine residual. The breakpoint is critically dependant on the pH of the system. As pointed out by Morris (1970), the occurrence of reactions giving rise to the breakpoint is most rapid in the pH range of 7.0 to 7.5. At greater and lower pH values, it becomes slower and less distinct; for example, at pH values of <6 or >9 the concept of breakpoint is not significant. In the pH range of 7.0 to 7.5, the breakpoint is about half developed within 10 min at 15 to 20°C and is then substantially completed within about 2 h.

Water samples to be chlorinated are recommended to be filtered prior to exposure to chlorine. It has been observed that removal of reductants such as chemically oxidisable organic matter, ammoniated compounds, and so forth, in the form of suspended matter from the water sample reduced the chlorine requirement for disinfection.

2.3.2 How Does Chlorine Kill Microorganisms?

Chlorine, when added to water, produces free chlorine, which consists of two chemical species: hypochlorous acid or HOCl, which is electrically neutral, and the hypochlorite ion or OCl$^-$, which is electrically negative. Hypochlorous acid is more reactive than the hypochlorite ion and is a stronger oxidant than the latter. It is evident that the ratio of both the species in water (equilibrium concentration) is an important factor that in turn is determined by the pH. Therefore, the speed and efficacy of water disinfection by chlorine will be affected by the pH of water being treated along with other oxidisable organic pollutants. Several bacteria and viruses are found to be easy targets of chlorination over a wide range of pH. Microbes such as *Giardia* are found to be exceptions to this rule and are much more resistant than most of the viruses and bacteria. Yet another reason for maintaining high concentration of hypochlorous acid during water treatment can be attributed to the fact that microorganisms carry a net negative charge on their surface. These surfaces are more readily penetrated by the uncharged, electrically neutral hypochlorous acid than the negatively charged hypochlorite ion. The microbes are killed or rendered incapable of reproducing due to the destruction of metabolic enzymes of the microbe or physical damage of the microbial DNA. Literature reports several probabilities for the destruction of microbes by chlorine, some of which are reported in Table 2.1.

2.3.3 Factors Affecting Chlorine Disinfection

For any disinfection technique to be effective there can be several contributing factors that have a direct impact on its efficacy. Type of source water, level of contamination, and type of microorganism, especially its cell wall structure, are some of the major factors. Apart from these, discussed below are some critical parameters related to the disinfectant being employed, particularly chlorine.

TABLE 2.1

Chlorine's Mechanisms of Action on Destruction of Microorganisms

Serial No.	Probable Mechanism of Action	Reference
1	Permeability of uncharged chlorine species.	Skvortsova and Lebedeva (1973), Kaminski et al. (1976), Dennis (1977)
2	Damage to the cell nucleus.	Chang (1944)
3	Multiple injuries to the cell surface.	Rahn (1945)
4	Causes the leakage of cytoplasmic material and inhibits the biochemical activities that are associated with the bacterial cell membrane, inhibiting both oxygen uptake and oxidative phosphorylation.	Venkobachar (1975), Venkobachar et al. (1977)
5	Causes certain bacteria and yeast to release protein or nucleic acid or their precursors and affects the uptake and retention of potassium by these same microorganisms.	Haas (1978)
6	Inhibition of the mechanism of glucose oxidation.	Green and Stumpf (1946)
7	Affects the respiration of bacteria as well as the rate of synthesis of protein and DNA.	Haas (1978)
8	Affects the nucleic acids or physically damages DNA.	Bocharov (1970), Bocharov and Kulikovskii (1971), Fetner (1962), Rosenkranz (1973), Shih and Lederberg (1976a, 1976b)
9	Extensive destruction of metabolic enzyme.	Chang (1971)
10	Initially, lethal damage is caused to the viral genome; it was observed that the capsid protein was affected only after the virus was inactivated.	Olivieri et al. (1975)

2.3.3.1 Concentration and Contact Time

In an attempt to establish more structured operating criteria for water treatment disinfection, the concentration multiplied by time (CXT) concept came into use in 1980. CXT is the final free chlorine concentration (mg/l) multiplied by minimum contact time (minutes). Based on the work of several researchers, CXT values offer water operators guidance in computing an effective combination of chlorine concentration and chlorine contact time required to achieve disinfection of water at a given temperature. The CXT formula demonstrates that if an operator chooses to decrease the chlorine concentration, the required contact time must be lengthened. Similarly, as higher-strength chlorine solutions are used, contact times may be reduced (Connell, 1996).

Contact time (CT) is usually expressed in mg-minutes per litre. CT is used widely and has been adopted by USEPA as a standard for specifying the requirements of chemical disinfection in water treatment. Generally chlorine has been regarded to be ineffective against *C. parvam* oocysts and the CT value required is very high, in the range of thousands of mg-minutes per litre. The CT values are dependent on the temperature and pH of water being treated and typical values for 4 log inactivation of viruses by free chlorine at a temperature of 0.5°C ranges from 12 (pH 6 to 9) to 90 (pH 10). At a temperature higher than 25°C, the CT values range from 2 (pH 6 to 9) to 15 (pH 10) (www.ce.pdx.edu).

2.3.4 Methods of Chlorine Treatment

Chlorine added to drinking water to destroy pathogenic organisms can be applied in several forms: elemental chlorine (chlorine gas), sodium hypochlorite solution (bleach), and dry calcium hypochlorite. While any of these forms of chlorine can effectively disinfect drinking water, each has distinct advantages and limitations for particular applications. Almost all water treatment systems that disinfect water use some type of chlorine-based process, either alone or in combination with other disinfectants.

2.3.4.1 Sodium Hypochlorite (Bleach)

Ordinary household bleach contains around 5 to 6% sodium hypochlorite (NaOCl) and can be used to purify water if it contains no other active ingredients, scents, or colourings. Bleach is far from an ideal source due to its bulkiness (only 5% active ingredient), and the instability over time of the chlorine content in bleach. Chlorine loss is further increased by agitation or exposure to air. Nevertheless, this may be the only chemical means available to purify water, and it is far better than nothing at all.

2.3.4.2 Calcium Hypochlorite (Bleaching Powder or Chlorinated Lime)

Calcium hypochlorite is another chlorinating chemical used primarily in smaller applications. It is a white, dry solid containing approximately

65% chlorine, and is commercially available in granular and tablet forms. Commonly called Cal-hypo and also known as high-test hypochlorite (HTH), calcium hypochlorite was patented in 1799 and called bleaching powder. It comes in a large granular form or as 1- or 3-inch tablets and is produced by passing chlorine gas over slaked lime. The resulting powder or granules provide 65 to 70% available chlorine. Cal-hypo will support combustion, and one needs to avoid mixing it with acids, ammonia, soda pop, oil, trichlor, or just about anything but water. Mixing with organics will cause a fire. Because it is slow to dissolve, it should be either used in a feeder or predissolved in water (around 12% by weight solution) and then added as a liquid. Cal-hypo can temporarily cloud the water, because the calcium takes a long time to dissolve completely. It can also cause calcium scaling and deposits on surfaces and in water circulating equipment. Cal-hypo tends to increase the water hardness level quickly and has a high pH of 11.8. Neutralising the pH requires approximately 12 m^3 of hydrochloric acid per 0.45 kg of cal-hypo.

This is the most effective method of chlorine treatment in the field. The U.S. military and most aid agencies also use HTH to treat their water, through a test kit, rather than use of superchlorination, that is, addition of an extra dose of chlorine, to ensure enough chlorine is added. This is preferable for large-scale systems, as the residual chlorine will prevent recontamination. Usually bulk water treatment plants first dilute to HTH to make a 1% working solution at the rate of 14 g HTH per litre of water. When test kits are available, they usually follow the WHO standard, that is, a residual chlorine level of 0.2 to 0.5 mg/l after 30 min of contact time. This may require as much as 5 mg/l of chlorine to be added to the raw water.

2.3.4.3 Chloramines

Combining ammonia (NH_3) with chlorine (Cl_2) to form chloramines for the treatment of drinking water has been called combined residual chlorination, chloramination, or the chloramine treatment process. Objectives of this water treatment are to provide disinfecting residual that is more persistent than free chlorine in distribution systems and to reduce the unpleasant tastes and odours that are associated with the formation of chlorophenolic compounds (Symons et al., 1977). Thus, this process utilises the formation of monochloramine (NH_2Cl) as indicated in the following reaction:

$$NH_3 + HOCl = NH_2Cl + H_2O \qquad (2.1)$$

The production of monochloramine is optimised by a pH range of 7 to 8 and a chlorine-to-ammonia ratio of 5:1 by weight or less (Symons et al., 1977). White (1972) states that the preferred ratio is 3:1 by weight. At higher chlorine-to-ammonia ratios or at lower pH values, dichloramine ($NHCl_2$) and trichloramine (NCl₃), also called nitrogen trichloride, are formed according to the following reactions:

$$NH_2Cl + HOCl = NHCl_2 + H_2O \qquad (2.2)$$

$$NHCl_2 + HOCl = NCl_3 + H_2O \qquad (2.3)$$

These and organic chloramines are produced during the chlorination of water containing ammonia or organic amines. Their presence may contribute to the unpleasant taste and odour problems in the finished water (Symons et al., 1977). Ammonia may be added to the water before (preammoniation) or after (postammoniation) addition of chlorine. Preammoniation can prevent the formation of tastes and odours that are caused by reaction of chlorine with phenols and other substances. However, according to White (1972), postammoniation is the most often used in the ammonia-chlorine water treatment process.

When chlorine is added to waters containing ammonia, the breakpoint phenomenon becomes significant in the pH range of 6 to 9. Chlorine-to-ammonia molar ratios ranging from 0 to 1 result in monochloramine formation. At values greater than 1, dichloramine is formed. Being unstable, it decomposes. Thus, with the addition of chlorine the apparent chlorine residual decreases from a chlorine-to-ammonia molar ratio of 1, up to approximately 1.65, at which the breakpoint occurs, that is, after the ammonia has been converted principally to nitrogen and some nitrate. Chlorine that is added after the breakpoint exists as free chlorine, that is, hypochlorous acid and the hypochlorite ion (OCl-) (Morris, 1978; Wei, 1972; Wei and Morris, 1974).

The mechanism of action is essentially the same as that of hypochlorous acid on enzymes (Nusbaum, 1952). However, the changes are generally seen in enzymes that may not be involved in the inactivation of the organism by hypochlorous acid (Ingols et al., 1953). The effects of chloramine on nucleic acids or DNA of cells cannot be ruled out (Fetner, 1962; Shih and Lederberg, 1976a, 1976b). A "multiple hit" concept, that is, when there might be three or four "targets" or points of attack, and that perhaps all must be affected before there is death of the microbe, has also been proposed (Fair et al., 1948).

2.3.4.4 Chlorine Dioxide

Chlorine dioxide is an effective bactericide and virucide under the pH, temperature, and turbidity ranges (pH 7 to 10, 25 to 37°C, 0.5 to 10 nephelometric turbidity unit (NTU), depending on the source of raw water) that are expected in the treatment of potable water. It should be noted that USEPA (1978) had set an interim maximum contaminant level (MCL) of 1 mg/l on chlorine dioxide because of the unresolved questions on its long-term health effects (Symons et al., 1977). Chlorine dioxide (ClO₂) is generated on-site at water treatment facilities. Two principal methods are used for the preparation of aqueous solutions of chlorine dioxide for water treatment (Gall, 1978; Masschelein, 1967, 1979). The first is the reaction of sodium chlorite ($NaClO_2$) and chlorine (Cl_2) in acidic aqueous solution, primarily by the following equation:

$$HOCl + 2ClQ + H = 2ClO_2 + Cl + H_2O \qquad (2.4)$$

According to Masschelein (1969), optimal yields of chlorine dioxide (90 to 95% of the theoretical yield based on sodium chlorite) are obtained by mixing a strong sodium chlorite solution (300 g/l) with aqueous chlorine solution containing 2 to 3 g/l of Cl_2. The mixed solution is then usually passed through a reactor that is packed with inert turbulence-promoting material to give a contact time of 1 to several minutes for reaction to be completed. The resulting solution, containing 3 to 5 g/l of chlorine dioxide, may be dosed directly into the water to be treated or may be diluted to about 1 g/l for short-term intermediate storage before dosing. The other method of preparing chlorine dioxide for use in water treatment is acidification of strong sodium chlorite solution, usually with hydrochloric acid (HCl). The major reaction is:

$$5ClQ + 4H = 4ClO_2 + Cl + 2H_2O \qquad (2.5)$$

Chlorine dioxide characteristics are quite different from chlorine. In solution, it is a dissolved gas, which makes it largely unaffected by pH, but it is volatile and can be relatively easily stripped from the solution. Chlorine dioxide is also a strong disinfectant and is selective in its attack on organic materials. While chlorine dioxide does produce a residual, it is only rarely used for this purpose.

Ingols and Ridenour in 1948 reported that adsorption of chlorine dioxide on the cell wall with subsequent penetration into the cell where it reacts with enzymes containing sulphhydryl groups was responsible for cell death. Later, it was observed that abruptly inhibiting protein synthesis (Benarde et al., 1967) could also be responsible for the disinfection.

Thus, it is very clear that chlorination can be brought about by various means, each having its own advantages and disadvantages. These are described in Table 2.2.

Up until the late 1970s, chlorine was virtually the only disinfectant used to treat drinking water. Chlorine was considered an almost ideal disinfectant, based on its proven characteristics. Although chlorine continues to be the ideal choice for a water disinfectant, it reacts with the naturally occurring humic and fulvic acids in water to produce trihalomethanes (THMs), which are toxic in nature. Water containing THMs, if consumed, can have a lethal effect owing to their carcinogenicity. Identification of newer disinfection by-products has spurred a lot of research in this particular area and is dealt with in Section 2.7.5. This critical drawback of chlorine led to an extensive

TABLE 2.2

Merits and Demerits of Various Chlorine-Based Chemical Disinfection Methods

Serial No.	Type of Chlorine Disinfectant	Advantages and Ranges of Concentrations Used	Disadvantages
1	Chlorination	Effective against most known pathogens; good residual action; suitable for most water quality conditions; easy to monitor; economical.	Harmful by-product formation (THMs and brominated organic by-product); not effective against *Cryptosporidium*; requires transport and storage of chemicals.
2	Elemental chlorine	Lowest cost in this form and unlimited shelf life; 61 to 84% (American Water Works Association, 2000).	Hazardous and requires special handling; additional regulatory requirement; highly corrosive in the presence of moisture.
3	Sodium hypochlorite (bleach) formed by adding elemental chlorine to sodium hydroxide; contains 5 to 15% chlorine	Less hazardous and easier to handle than elemental chlorine; fewer training requirements and regulations than to elemental chlorine; 17 to 34% (American Water Works Association, 2000).	Limited shelf life; forms by-products like chlorate and chlorite; corrosive to some materials; more difficult to store than most chemical solutions; and higher cost than elemental chlorine.
4	Calcium hypochlorite (produced by passing chlorine gas over slaked lime); commonly called bleaching powder; contains 65 to 70% available chlorine	More stable than sodium hypochlorite; fewer training requirements and regulations than elemental chlorine; 1 to 9% (American Water Works Association, 2000).	Dry chemical requires more elaborate handling than sodium hypochlorite; precipitated solids can cause difficulty in chemical feeding; higher cost than elemental chlorine; fire or explosive hazard; potential to add by-products like chlorate and chlorite to water.

(Continued)

TABLE 2.2 (CONTINUED)

Merits and Demerits of Various Chlorine-Based Chemical Disinfection
Methods

Serial No.	Type of Chlorine Disinfectant	Advantages and Ranges of Concentrations Used	Disadvantages
5	On-site hypochlorite generation (using electrolytic cell and salt solution); produces weak solution of approximately 0.8% of chlorine	Minimal chemical storage and transport; for each pound of equivalent chlorine formed, 3 lb of salt and 2 kWh of power are required; the resultant hypochlorite solution has a concentration of 0.8% (www.water.siemens.com/.../On-site_Hypochlorite_Generation.pdf).	More complex; higher maintenance and expertise; high capital and operating cost; by-products generated may be difficult to monitor and control; additional maintenance of electrolytic cell.
6	Chloramines	Reduced formation of THM and brominated by-products; more stable residual than free chlorine; excellent secondary disinfectant and prevents biofilm growth; lower taste and odour than free chlorine; 2 to 29% (American Water Works Association, 2000).	Weak disinfectant and oxidant; requires shipment and handling of ammonia and chlorinated compounds; ammonia is hazardous to human and aquatic life; longer contact times of 5 to 30 min.
7	Chlorine dioxide	Effective against *Cryptosporidium* and *Giardia* species compared to chlorine; does not form chlorinated by-products like THM and HAA; more effective in treating some taste and odour problems than chlorine; 6 to 8% (American Water Works Association, 2000).	Inorganic by-products are formed (chlorite and chlorate); highly volatile residuals; requires on-site generation equipment and handling of chemicals (chlorine and sodium chlorite); high technical competence to operate and monitor equipment, product, and residuals; may pose odour and taste problems; high material (chlorite chemical) cost and hence operating cost.

search for alternative disinfectants that did not produce harmful disinfection by-products. Moreover, drinking water providers have faced an array of new challenges when chlorine is used alone, which include treating resistant pathogens such as *Giardia* and *Cryptosporidium*, new environmental and safety regulations regarding the residual chlorination, and strengthening operational safety at the treatment facilities during the handling of chlorine chemicals.

More recently, to meet these new challenges, a water system must be designed with unique disinfection approaches to match each system's characteristics and source water quality. While chlorination remains the most commonly used disinfection method by far, water systems may use alternative disinfectants, including chloramines, chlorine dioxide, ozone, and ultraviolet radiation, or their combination. No single disinfection method is right for all circumstances, and in fact, water systems may use a variety of methods to meet overall disinfection goals at the treatment plant, and to provide residual protection throughout the distribution system.

The sections below describe various disinfection technologies other than chlorination, and discuss the major advantages and limitations associated with each of these disinfection chemicals and techniques.

2.4 Other Chemical Methods

2.4.1 Iodine

After World War II, iodine emerged as a replacement for halazone (halogen and ozone) tablets for water disinfection, as it was found to be superior to chlorine for treating small batches of water. In general, iodine was found to be less sensitive to pH and organic content of water and effective at lower doses. The salient drawback was the allergic reactions exhibited by individuals who were sensitive to it. Therefore, long-term usage of iodine as a disinfectant has always been a concern for experts in the field. The recommended dietary allowance (RDA) of iodine varies for infants, children, and pregnant women. Generally, it is 150 μg/day for adults. In areas using iodine as a disinfectant, use of iodised salt may be discouraged so that the RDA is not overshot.

Water disinfection is very important, and especially critical for soldiers, hikers, and others who are likely to drink water from any outdoor source. Iodine has been the disinfectant of choice in such cases. Iodine tablets, as recommended by the manufacturer, are added to water to ensure potability. Although this successfully inactivates most of the indicator microorganisms found in water, such as *Escherichia coli* and faecal coliforms, protozoan parasites like *Cryptosporidium* and *Giardia* are much more difficult to kill. Iodine is normally used in doses of 8 ppm to treat clear water over a 10 min contact time. The effectiveness of this dose has been shown in numerous studies. Cloudy or turbid water needs twice as much iodine or twice as much contact time. In cold water (below 41°F or 5°C) the dose or the contact time must also be doubled. In any case, doubling the treatment time will allow the use of half as much iodine. Prefiltration of the water will also reduce the required iodine dosage. These doses are calculated in terms of that required to remove all pathogens (other than *Cryptosporidium*) from the water. Of these, *Giardia* cysts are the hardest to kill, and thus require a high level of iodine. If the cysts are filtered out with a microfilter, only 0.5 ppm of iodine is needed to successfully treat the resulting water.

In a study conducted by Ellis and Van Vree (1989), 2 mg/l of iodine has been reported to be comparable with 1 mg/l of chlorine in its efficacy against faecal indicator bacteria at low pH (<5) and turbidity with contact time of 30 min. However, increasing pH and turbidity were found to decrease the disinfecting capabilities of both chlorine and iodine (Ellis and Van Vree, 1989). Disinfection of water containing resistant cysts of *Giardia* is not as easy to tackle. In fact, the dynamics of disinfection have not been studied satisfactorily, particularly with respect to these resistant organisms, which are human pathogens. Scanty work has been reported in this aspect. In one interesting study using *Giardia muris* as a model for the human pathogen, iodine inactivation of cysts over a wide range of temperature and iodine concentrations revealed that the manufacturer's recommendation of iodine dosage failed to successfully inactivate these cysts, and new recommendations of iodine dosage were required to ensure complete disinfection (Fraker et al., 1992). Yet another study clearly points out the inability of iodine, and hence the necessity of alternate disinfection methods to inactivate *Cryptosporidium* oocycts. In this case, *Giardia* cysts and *Cryptosporidium* oocysts were exposed to iodine, according to the manufacturer's instructions (two tablets/l = 13–18 mg/l for 20 min). Compared to *Giardia* cysts, the *Cryptosporidium* oocycts were difficult to inactivate. Only 10% were inactivated after a 20 min exposure, and only 66 to 81% of oocysts were inactivated after 240 min (Gerba, Johnson, and Hasan 1997). Although cysts are a challenge yet to be overcome, iodine's performance as a viral disinfectant appears to be quite good. Literature reports the inactivation studies carried out on the model MS2 coliphage virus. Although 99.99% disinfection was achieved at an iodine dosage of less than 5 mg/l within 3 min of contact time, an apparent rebound was observed in the presence of beef protein extracts. This was an interesting finding and was attributed to the reaction between the oxidised iodine residual and the beef extract proteins in solution (Brion and Silverstein, 1999). Iodine disinfects viruses by causing conformational changes in their protein coat. Being proteins in nature MS2 virions have an isoelectric focusing point at pH 3.9. Reaction with iodine has been shown to cause a shift in the isoelectric pH to basic values. This suggests that iodine causes changes in the charge distribution characteristics of the protein coat (Brion and Silverstein, 1999).

In general, compared to chlorine, iodine is much more effective in treating poor quality water, such as that containing sludge. An iodine concentration ranging from 1 to 10 mg/l was utilised to disinfect three types of stream waters containing stream sediments, digested sludge, and raw sludge, respectively, at different pH and turbidity ranges. The efficiency of iodine assessed by percentage *E. coli* inactivated was found to be better than chlorine, especially at higher temperatures and pH values (Ellis, Cotton, and Khowaja, 1993).

Water treated with iodine, which may have any objectionable taste, can be removed by treating the water with vitamin C (ascorbic acid), but it must be added after the water has been disinfected with iodine. Ascorbic acid reacts with iodine to form a complex, thereby enabling the removal of the objectionable taste. Flavoured beverages containing vitamin C will accomplish the same thing. Sodium thiosulphate can also be used to combine with free iodine, and either of these chemicals will also help remove the taste of chlorine as well. Usually, elemental iodine cannot be tasted below 1 ppm, and below 2 ppm the taste is not objectionable. Iodine ions have an even higher taste threshold of 5 ppm. It is interesting to note that removing the iodine taste based on the complexation of iodine does not reduce the required limitation of the dose of iodine ingested into the body.

The mechanism of action of iodine has been researched a lot, and for bacteria it is attributed to a reaction with vital amino acids in the proteins (Chambers et al., 1952; Hsu, 1964; Hsu, Nomura, and Kruse 1966). The mode of action of iodine in cyst penetration has not been studied (Brammer, 1963; Hsu, 1964). It is reported that biomolecular reaction with a single iodine molecule and clumping of virions played a role in resistance to disinfection (Berg et al., 1964), but the hydrated cationic iodine species attacks the nucleotide base (Allen and Keefer, 1955; Bell and Gelles, 1951; Hsu, 1964; Hughes, 1957).

Research on iodine as a water disinfectant continues to give better insight and reveals newer perspectives not known earlier. It is still regarded as a boon to the traveller, as it can be easily carried in transit and applied to water before consumption to prevent diarrhoea. Recently, Heiner et al. demonstrated the effectiveness of 10% povidone-iodine (PVI) in disinfecting field water within 15 min of contact time. *E. coli*, the microbe often associated with traveller's diarrhoea, was successfully disinfected. They claim that the same concentration of iodine may be enough to disinfect other pathogenic microorganisms. This clearly brings out the usefulness of simple techniques for water disinfection (Heiner et al., 2010).

Thus, iodine has many features that are comparable to free chlorine and bromine as a water disinfectant, but iodamines are not formed. Free iodine is an effective bactericide over a relatively wide range of pH. Field studies on small public water systems have shown that low levels of 0.5 to 1 mg/l of free iodine can be maintained in distribution systems, and that the magnitude of residual is sufficient to produce safe drinking water with no adverse effects on human health. Like other halogens, the effectiveness of iodine against bacteria and cysts is significantly reduced by high pH, but unlike bromine and chlorine, it is much more effective against viruses because of the enhanced iodination of tyrosine. Currently, its use is restricted primarily to emergency disinfection of field water supplies because of its high cost and because it is difficult to apply to large systems. The possible adverse health effects of increased iodide intake for susceptible individuals in the population must also be considered. Studies should be conducted to determine the consequences for human health of the long-term consumption of iodine in drinking water, with special regard for more susceptible subgroups of the population.

2.4.2 Silver

Silver has been suggested by some (Just and Szniolis, 1936; Newton and Jones, 1949; Yasinskii and Kuznetsova, 1973) for water treatment. Its use was initially out of favour due to USEPA's establishment of a 50-ppb maximum contaminant level limit on silver in drinking water. This limit was set to avoid argyrosis, a cosmetic blue/grey staining of the skin, eyes, and mucous membranes. As the disease requires a net accumulation of 1 g of silver in the body, it can be calculated that one could drink water treated at 50 ppb for 27 years before accumulating 1 g. Thus, to achieve acceptable disinfection in a reasonable time, concentrations exceeding the MCL of 0.05 mg/l were recommended. In spite of the general toxicity, ecotoxicity, and sparsely known metabolism of silver, WHO allows silver ions (Ag) up to 0.1 mg/l in drinking water disinfection. However, literature reports on the effects of silver on animal models clearly indicate a reevaluation of the present recommendations on the use of silver salts for disinfection of drinking water might be necessary. For example, studies on mice conducted by giving drinking water containing a threefold lower concentration of 0.03 mg/l silver ions as silver nitrate labelled with [110m]Ag for a period of 1 to 2 weeks revealed results that may have serious implications on humans. The silver concentrations analysed by gamma radiography showed highest concentrations of silver in the musculus soleus, cerebellum, spleen, duodenum, and myocardial muscle of the mice in rank order. The accumulation of silver into organs and tissues important in motor functions may be of relevance especially in emergency and catastrophe situations in which accurate motor functions may be critical (Kai, Tanski, and Hänninen, 2003).

Silver has only been proven to be effective against bacteria and protozoan cysts, though no specific studies were reported in the literature for their action against viruses. Silver can be used in the form of a silver salt, commonly silver nitrate, a colloidal suspension, or a bed of metallic silver. Electrolysis can also be used to add metallic silver to a solution. Some evidence has suggested that silver deposited on carbon block filters can kill pathogens without adding as much silver to the water. Avid research has led to innovative techniques to incorporate silver in water that includes nanoparticles. The last decade has witnessed extensive innovations in the arena of silver nanotechnology for water disinfection. Very recently, polymeric microspheres developed by anchoring silver nanoparticles (NPs) onto macroporous methacrylic acid copolymer beads have been tried for disinfection of water. They proved to be highly effective against both gram-positive and gram-negative bacteria tested, except spore-forming *Bacillus subtilis*, which showed a 99.9% reduction, against 100% in the former case. Moreover, there was no bacterial adsorption on the copolymer beads containing silver nanoparticles (Gangadharan et al., 2010).

Today, silver is regarded not only as an antimicrobial agent but also as an effective antiviral disinfectant. Recently, the association of silver nanoparticles with the bacterial cell surface of *Lactobacillus fermentum* (referred to as biogenic silver or bio-Ag^0) has been reported to exhibit antiviral properties. The microscale bacterial carrier matrix serves as a scaffold for Ag^0 particles, preventing aggregation during encapsulation. In this study, bio-Ag^0 immobilised in different microporous polyvinylidene fluoride (PVDF) membranes inactivated bacteriophages by slow release of Ag$^+$ from the membranes, showing the potential of this membrane technology for water disinfection on a small scale (De Gusseme et al., 2011). Yet another interesting study on the use of micron-scale graphene oxide (GO) nanosheets impregnated with silver nanoparticles as bactericidal agent for water disinfection indicates the extensive work in this arena of water disinfection. These paper-like Ag-NP/GO composite materials exhibited strong antibacterial action against *Escherichia coli* and *Staphylococcus aureus*, which were used as model strains of gram-negative and gram-positive bacteria, respectively. These were fabricated owing to convenient structure characterisation and antibacterial tests, enabling them to be easily deposited on porous ceramic membranes during water filtration, and making them a promising biocidal material for water disinfection (Bao et al., 2011).

Control of biofouling and its negative effects on process performance of water systems is often a serious operational challenge in all the water sectors. Molecularly capped silver nanoparticles (Ag-MCNPs) used as a pretreatment strategy for controlling biofilm development in aqueous suspensions using the model organism *Pseudomonas aeruginosa* retarded biofilm formation up to 60% and more. Although further studies are warranted, the results are indeed very promising (Dror-Ehre et al., 2010).

Research has also been directed toward newer techniques for determination of silver in drinking water that are quicker, accurate, and easy to use in varied scenarios, including spacecrafts! At this point it is very interesting to note that silver is added, as biocide, to drinking water for spacecrafts. Bruzzoniti et al. (2010) report a flow injection method for determination of silver that is based on a reduction reaction with sodium borohydride, which leads to the formation of a colloidal species that is monitored at a wavelength of 390 nm. This technique effectively determines silver in the range of 0.050 to 5.0 mg/l with a minimum detectable concentration of 0.050 mg/l. The method optimised was applied to a drinking water sample provided for the launch with the Automated Transfer Vehicle (ATV) module Jules Verne to the International Space Station on March 9, 2008 (Bruzzoniti et al., 2010).

Silver inactivation of microbes has been reviewed a lot in literature, and it may be due to either one or a collective mechanism of several possibilities that exist. As early as 1952, it was observed that at low concentrations, silver adsorbed onto the bacterial surface just as silver tends to be adsorbed on other surfaces. This inhibits cellular respiration, which occurs on the cell surface (Zimmermann, 1952). Later, this was corroborated by Chang, who recognised the mechanism as a direct action of silver in the nonreversible formation of silver-sulphhydryl complexes that could not function as hydrogen carriers. The adsorbed silver ions must immobilise the dehydrogenation process because bacterial respiration takes place at the cell surface membrane. He considered silver to be bacteriostatic as well as bactericidal (Chang, 1970).

2.4.3 Potassium Permanganate

Potassium permanganate ($KMnO_4$) is a strong oxidising agent, which was first used as a municipal water treatment chemical by Sir Alexander Houston of the London Metropolitan Water Board in 1913. In the United States, it was used in Rochester, New York, in 1927 and in Buffalo, New York, in 1928. Since 1948, it has been used more widely in waterworks as an algaecide (Fitzgerald, 1964; Kemp et al., 1966), as an oxidant to control tastes and odours (Spicher and Skrinde, 1963; Welch, 1963), to remove iron and manganese from solution (American Water Works Association, 1971; Shull, 1962), and to a limited extent, as a disinfectant (Cleasby et al., 1964). The relatively limited information concerning disinfection with potassium permanganate was subject to criticism, because there were no studies of the effects of organic constituents of the medium (test system) or destruction of a variety of organisms, especially pathogens. Therefore, initial focus of research was directed to address some of these concerns. As early as 1950, studies conducted on Indian well waters revealed that a 10 ppm concentration of potassium permanganate was sufficient to kill *Vibrio cholerae* in 2 h, *Shigella flexneri* in 4 h, and *Salmonella typhi* in 24 h. However, higher doses of above 20 ppm were required to kill the typhoid bacteria if present in large numbers. However, it was concluded that the pink colour remaining in water up to 24 h after treatment of well waters with potassium permanganate could not be considered an indication of achievement of safe water (Banerjea, 1950). Studies on antibacterial properties of potassium permanganate continued and were found to be effective against a wide spectrum of microbes, such as *Legionella pneumophila* (Yahya et al., 1989).

As mentioned previously, potassium permanganate has always been widely used as an algaecide. However, critical studies giving insight on the mechanism of inactivation and effect of hardness of water only happened in the last decade. The effect of potassium permanganate as preoxidant for algae-laden source water revealed that potassium permanganate promotes the aggregation of algae cells, and this phenomenon was even more significant with the existence of a hardness-causing ion, calcium. It was also found to incorporate its reducing product, manganese dioxide (MnO_2), into algae floc, and increased its specific gravity, and therefore its settling velocity. In addition, permanganate was also found to induce the release of extracellular organic matters (EOMs) from algae cells (Chen and Yeh, 2005).

Crystalline potassium permanganate is highly soluble in water (2.83 g/100 g at 0°C). In waterworks, it is prepared usually as a dilute solution (1 to 4% by weight) and applied with a chemical metering pump (American Water Works Association, 1971). It also may be added as a solid using conventional dry-feed equipment. Reduction of the permanganate ion produces insoluble manganese oxide (Mn_3O_4) hydrates. To prevent distribution of turbid water or of water that will cause unsightly staining of plumbing fixtures, potassium permanganate most often is applied as a pretreatment that is followed by filtration. For example, addition of potassium permanganate to a finished water to maintain a residual in a distribution system is unacceptable because of the pink colour of the compound itself or the brown colour of the oxides. Potassium permanganate is a much weaker oxidising agent than the alternatives, more expensive, and leaves an objectionable pink or brown colour. If it must be used, 1 g/l would probably be sufficient against bacteria and viruses (no data are available on its effectiveness against protozoan cysts).

The mechanism of inactivation by potassium permanganate has been ascribed to oxidising compounds that are involved in essential cellular functions (Lund, 1963a,b, 1966).

2.4.4 Hydrogen Peroxide

Hydrogen peroxide (H_2O_2) is a strong oxidising agent that has been used for disinfection for more than a century. Its instability and the difficulty of preparing concentrated solutions have tended to limit its use. However, by 1950, electrochemical and other processes were developed to produce pure hydrogen peroxide in high concentration, which is known as stabilised hydrogen peroxide (Schumb, Satterfield, and Wentworth, 1955). This product has been subjected to increased study and application. It has been used to disinfect spacecraft (Wardle and Renninger, 1975), foods (Toledo, 1975), and contact lenses (Gasset et al., 1975; Spaulding et al., 1977). Although there has been some interest in using hydrogen peroxide as a disinfectant for wastewater, it has been used more for control of bulking in the activated sludge waste treatment process (Sezgin et al., 1978). Its use in drinking water disinfection initially appears minimal.

2.4.4.1 Production and Application

Commercially produced hydrogen peroxide is available in aqueous solution, usually ranging from 30 to 90%. It is stabilised during manufacture by addition of such compounds as sodium pyrophosphate ($Na_4P_2O_7$). As the concentration of hydrogen peroxide is decreased, the concentration of stabilisers is typically increased. Experience in the waterworks industry using hydrogen peroxide is limited. However, hydrogen peroxide is widely used as a bleaching agent in making cotton textiles or in wood pulping (Chadwick and Hoh, 1966). Presumably, a concentrated solution would be diluted and applied with a chemical metering pump. Careful attention to safety in handling would be required because of the possibility of fire or explosion.

Because of its relatively high cost and the high concentrations (>100 to 150 ppm) that are required to achieve disinfection in a reasonable time, hydrogen peroxide is not a generally satisfactory disinfectant for drinking water. Reports on the use of hydrogen peroxide as a sole disinfectant of water for potability are scanty. However, some work was carried out by the authors that is described in detail in Chapter 5. Hydrogen peroxide has been used in combination with other chemical disinfectants, especially ozone, as an advanced oxidation process. It has been established now with substantial evidence that hydrogen peroxide works very well synergistically and disinfects water by production and action of powerful oxidising radicals. This has also been described at length in Chapter 4.

It is well established that the hydrogen peroxide molecule itself was not responsible for the action but, rather, that the free hydroxyl radical (HO˙) that it produced was the specific inactivating agent (Spaulding et al., 1977). The presence of free radicals was important, but whether sufficient catalysing metal ions were available either from the tap water that was used or from the bacterial cells themselves was not really understood (Yoshpe-Purer and Eylan, 1968).

Disinfections studies on *E. coli* reveal that concentrations above 100 ppm were required for effective inactivation, but the contact time was far too long compared to other disinfection methods. Below 40 ppm, the inactivation was practically ineffective (Labas et al., 2008). In addition to bacterial inactivation, hydrogen peroxide finds its importance due to its ability to prevent biofilm formation. Compared to disinfectants like chlorine, chloramine, and ozone, hydrogen peroxide has a longer residual effect and therefore controls biofilm growth for a long period, at least up to the residual concentration in the system (Momba et al., 1998). In spite of its effectiveness, hydrogen peroxide is not very stable, which also dampens its performance. Efforts to tackle this issue lead to a promising solution in the form of a stabilised H_2O_2-based compound, generally called hydrogen peroxide plus (HPP). Its effectiveness in disinfecting water containing pathogen microbes was recently evaluated. It is remarkable to note that faecal coliforms were the most sensitive to treatment, followed by somatic coliphages and F+ bacteriophages. These counts were reduced to almost 99% within a contact time of 35 min, proving its usefulness and feasibility in treating small systems in small communities and private houses, compared to chlorine (Ronen et al., 2010).

2.4.5 Bromine

Bromine was first applied to water as a disinfectant in the form of liquid bromine (Br2) (Henderson, 1935), but it can also be added as bromine chloride gas (BrCl) (Mills, 1975) or from a solid brominated ion exchange resin (Mills, 1969). Oxidation of bromide (Br$^-$) to bromine can also be accomplished either chemically or electrochemically. Oxidation with aqueous chlorine gives either bromine or hypobromous acid (HOBr), depending on the ratio of chlorine to bromide. Both bromine (Liebhafsky, 1934) and bromine chloride (Kanyaev and Shilov, 1940) hydrolyse to hypobromous acid.

Laboratory tests show that bromine is an effective bactericide and virucide. It is more effective than chlorine in the presence of ammonia. As a cysticide, it is highly active. Bromine is active over a relatively wide range of pH, and it retains some of its effectiveness as hypobromous acid to above pH 9. The major disadvantage of bromine as a disinfectant is its reactivity with ammonia or other amines that may seriously limit its effectiveness under conditions that are encountered in the treatment of drinking water. Data on the effectiveness of bromine against bacteria were complicated by this reactivity and the lack of characterisation of the residual species in disinfection studies. Further research was gradually directed to quantify the effectiveness of the various species of bromine and the bromamines, both organic and inorganic, against bacteria, viruses, and cysts. The reactivity of bromine and the dosages that are required to maintain effective residual in natural water systems were also investigated. The effectiveness of various technologies for application of bromine or bromine chloride to intended drinking water was also evaluated (Johnson and Overby, 1971; Taylor and Butler, 1982).

Bromine inactivates microorganisms by moving to the vital site and affecting the mass transport, and it reacts by oxidation of that site. In case of viruses it is believed to affect the protein coat (Olivieri et al., 1975).

2.4.6 Ferrate

Ferrates are salts of ferric acid (R.FeO$_4$, where *R* is alkaline metal) in which iron is hexavalent. Fremy (1841) first synthesised potassium ferrate (K$_2$FeO$_4$) in the mid-nineteenth century. Since then, a wide variety of metallic salts have been prepared. However, only a few of the preparations yield ferrates of sufficient purity and stability for use in the treatment of water. Ferrates are strong oxidising agents that have redox potentials of –2.2 and 0.7 V in acid and base, respectively (Wood, 1958). Aqueous solutions of potassium ferrate are unstable and decompose to yield oxygen (O$_2$), hydroxide (OH$^-$), and insoluble hydrous iron oxide [FeO(OH)]: The hydrous iron oxide is a coagulant that is commonly used in water treatment. Initial ferrate concentration, pH, temperature, and the surface character of the resulting hydrous iron oxide affect the rate of decomposition (Schreyer and Ockerman, 1951; Wagner et al., 1952; Wood, 1958). Ferrate solutions are most stable in a strong base (>3 M or at pH 10 to 11). Schreyer and Ockerman (1951) reported that dilute aqueous solutions of ferrate are more stable than concentrated solutions. The presence of other inorganic ions also affects ferrate stability in aqueous solutions. Ferrate reacts rapidly with reducing agents in solutions

(Murmann, 1974). It will also oxidise ammonia (NH_3). The rates of this reaction increase with pH (optimum pH range 9.5–11.2), molar ratio of ferrate to ammonia, and temperature (Strong, 1973). The oxidation of ammonia by ferrate, below pH 9, is markedly slower than that observed for chlorine. Some information is available on the reaction of ferrates with various organic compounds (Audette, Quail, and Smith, 1971; Becarud, 1966; Zhdanov and Pustovarova, 1967), which could be effectively used to understand the action of ferrate in disinfection in the presence of other chemical contaminants.

The disinfecting properties of ferrate (VI) were first observed by Robinson and Murmann (1975) when they found that ferrate could effectively kill laboratory cultures of wild-type and recombinant cultures of *Pseudomonas* over a concentration range of 0–50 ppm. Further studies (Gilbert et al., 1976) have demonstrated its efficacy in destroying *E. coli* to 99.9% when the dosage was 6 mg/l and the contact time 7 min. Higher contact times were required if dosage was to be reduced. Viruses have always been a challenge to destroy, as they can exist at low concentrations for prolonged periods in water. Ferrate has been shown to inactivate viruses successfully (Kazama, 1995; Shink and Waite, 1980). The virus used was f2 coliphage, which belongs to the group Picornaviridae, and is frequently found in sewage and is more resistant to chlorination than polivirus (which also belongs to the same group). The results demonstrated that potassium ferrate rapidly inactivated f2 coliphage. Moreover, the results showed that K_2FeO_4 was more effective than chlorine and bromine, and no K_2FeO_4 residual was found in the treated water. In contrast, chlorine and bromine normally leave residuals in the treated water.

In one study, for example, THM formation potential (THMFP; 4 h) in treated water with ferrate was 75% less than that with chlorination (Waite and Gray, 1984).

Viral inactivation by ferrates was researched almost four decades ago, and it was always attributed to the disruption of capsid and subsequent loss of RNA, along with its extensive breaking down after release (Boeye and Van Elsen, 1967; Maizel, Phillips, and Summers, 1967; Van Elsen and Boeye, 1966).

2.4.7 Water Disinfection by Zero-Valent Iron Nanoparticles

Nanotechnology is a fast-growing area that has incited substantial attention in areas such as energy, electronics, drug delivery, and medical diagnostics, and also extended to the water treatment field. It has been articulated that nanotechnology has tremendous potential to transform the conventional water treatment methodologies (USEPA, 2007). Nanoparticles, especially zero-valent iron particles, have been extensively researched and applied in the remediation of organic and inorganic pollutants. Typically, iron exists in the environment as iron (II) and iron (III) oxides. Therefore, the zero-valent form has to be manufactured for its intended application. The usefulness of the zero-valent iron particle is owed to its electron donating property under ambient conditions, making it an attractive candidate for remediation (Stumm and Morgan, 1996).

Permeable reactive barriers (PRBs) are groundwater treatment structures that essentially have an engineered iron wall through which the water to be treated passes passively. This results in the contaminants being precipitated, adsorbed, or transformed when they come into contact with the iron wall. These have been used extensively in the United States since the 1990s (Reynolds et al., 1990). Although initial treatments used granular zero-valent iron, the nanoscale iron may be regarded as an alternative to the conventional PRBs (Li, Elliott, and Zhang, 2006). Numerous pollutants, such as perchloroethane, trichloroethane, carbon tetrachloride, and pesticides like lindane and DDT, along with heavy metals such as lead and chromium, have been effectively treated with iron nanoparticles (Alowitz and Sherer, 2002; Cao et al., 2005; He and Zhao, 2005; Kanel et al., 2005; Sohn et al., 2006; Xu et al., 2005). Some reports on the use of nanoscale iron appear in literature, especially in the inactivation of viruses. In an interesting study, iron-coated sand was investigated for disinfecting water, and it was found to adsorb viruses leading to its inactivation, which was primarily due to electrostatic interaction between the nano iron and the viral capsid (Ryan et al., 2002; You et al., 2005).

As far as antimicrobial action is concerned, nanomaterials such a silver, chitosan, and titanium dioxide also appear to have been researched and applied in water treatment, among which silver nanoparticles have been most widely employed. Several home water purification systems also use silver-coated membranes to enhance the antimicrobial properties (Maynard, 2007). Similarly, nanoscale chitosan, which is typically found in personal care products, antimicrobials in agriculture, and biomedical products, has potential as a water disinfectant owing to its cell wall disrupting properties and ability to chelate trace metals required by bacteria for their survival (No et al., 2002; Rabea et al., 2003). Water disinfection using titanium dioxide is well established and has been discussed at length in previous sections. It is known that TiO_2-based solar disinfection has very good water disinfection properties and is very effective in remote rural areas that do not have access to electricity. However, this usually takes a very long time, and the process can be accelerated by using metals such as silver that act synergistically, bringing about disinfection (Reddy et al., 2007; Page et al., 2007). Since hybrid processes have always proven to be beneficial, the combination of nanomaterial with existing water treatment techniques appears to be favourable. Nanoscale titanium dioxide (TiO_2) can be combined with UV disinfection systems to bring down the UV exposure time and effectively enhance the inactivation of microbes. This has been reported for pathogens such as *Cryptosporidium* and *Giardia*, where their cysts forms require too long an exposure time with conventional UV methods (Yates et al., 2006). Like any other water disinfection system, nanomaterial also has a plethora of limitations that may call for a very careful usage in water disinfection. If they are incorporated in the sample water forming a slurry, they will have to be separated downstream of the treatment train to retrieve and reuse it given the cost of its manufacture. Their presence in treated water is speculated to be hazardous since some of the nanomaterial used for water disinfection, like zinc oxide (ZnO), appear to reduce the viability of human T cells that play an important role in maintaining immunity of an individual (Reddy et al., 2007). Moreover, lack of residual action while using this technique also warrants the addition of secondary disinfectants to maintain the water quality until it reaches the end user. Although, on the positive side, nanoparticles are proving to be effective antimicrobials in water disinfection on account of their very small size and huge surface area, further research requirements in designing better technologies to retain them in an attempt to not only reduce loss of material and hence cost, but also to prevent human and environmental hazards if they escape into the finished product, is the need of the hour (Li et al., 2008).

2.5 Chemical Disinfection Treatments Requiring Electricity

2.5.1 Ozone

Initially, ozone was used extensively in Europe to purify water. Ozone, a molecule composed of three atoms of oxygen rather than two, is formed by exposing air or oxygen to a high-voltage electric arc. Ozone is much more effective as a disinfectant than chlorine, but no residual levels of disinfectant exist after ozone turns back into O_2. Ozone was always expected to see increased use in the United States as a way to avoid the production of trihalomethanes due to the usage of chlorine. While ozone did break down organic molecules, sometimes this was a disadvantage, as ozone treatment produced higher levels of smaller (lower molecular weight) molecules that provide an energy source for microorganisms or higher levels of toxicity. If no residual disinfectant is present (as would happen if ozone were used as the only treatment method), these microorganisms cause the water quality to deteriorate rapidly during the storage and distribution. Ozone also changes the surface charges of the dissolved organics and colloidally suspended particles. This causes microflocculation of the dissolved organics and coagulation of the colloidal particles.

Ozone (O3) is generated on-site at water treatment facilities by passing dry oxygen or air through a system of high-voltage electrodes. Ozone is one of the strongest oxidants and disinfectants available. Its high reactivity and low solubility, however, make it difficult to apply and control. Contact chambers are fully contained, and nonabsorbed ozone must be destroyed or recycled prior to its release into the atmosphere to avoid corrosive and toxic conditions. Ozone is particularly effective against spores and cysts. Extensive studies on various factors affecting *Giardia muris* cyst inactivation by ozone are reported. These encompass factors like temperature, turbidity, pH, ozone dose, and contact time. It was found that residual and utilised ozone both had important influences in *G. muris* cyst inactivation. It was more difficult to achieve 2 or 3 log inactivation of *G. muris* cysts in the natural waters at 22°C than at 5°C (Labatuik, Belosevic, and Finch, 1994). *Encephalitozoon intestinalis*, a microsporidian pathogen for humans and animals that is detected in surface water, has been listed as a major potential emerging waterborne pathogen by the USEPA. Ozone has been very effective in rendering these spores inactive, and its CXT values were an order of magnitude lower than the CXT values of chlorine in a comparative study (John et al., 2005).

Ozone has several advantages, such as being one of the strongest oxidants, not producing chlorinated THMs or HAAs, and being effective against *Cryptosporidium*, to name some. On the other hand, its limitations are also numerous. Its process operation and maintenance requires a high level of technical competence, it provides no protective residual, and it reacts with bromine-containing compounds and forms brominated by-products like bromate and brominated organics, and forms nonhalogenated by-products such as ketenes, organic acids, and aldehydes, which could be more toxic than the original contaminants. Ozone breaks down more complex organic matter, and smaller compounds can enhance microbial regrowth in distribution systems and increase disinfectant by-product (DBP) formation during secondary disinfection processes. Its higher operating and capital costs than chlorination are also a major limitation. Difficulty to control and monitor, particularly under variable microbial and organic load conditions, also adds to the list of ozone's demerits as a disinfectant.

The mechanism of action of ozone on microbes has been studied extensively. The primary attack of ozone occurs on the double bonds of the fatty acids (lipid layers) in the cell wall and membrane, and there is a consequent change in the cell wall permeability and cell contents leak out, causing death (Scott and Lesher, 1963; Smith, 1967). Attack on nucleic acids inside the cells is yet another mechanism of cell death. Thymine was more sensitive to ozone attack than were cytosine and uracil (Prat, Nofre, and Cier, 1968). In the case of viruses, a complete loss of viral proteins was responsible for the death of viruses (Riesser et al., 1977) as a result of an ozone attack.

Various chemical disinfection techniques have been successfully adopted to disinfect water and are reported in Table 2.3.

2.6 Coagulation/Flocculation Agents as Pretreatment

This can be regarded as a pretreatment to reduce chemical dosage. While flocculation does not kill pathogens, it will reduce their levels along with removing particles that could shield the pathogens from chemical destruction, and organic matter that could tie up chlorine added for purification as

TABLE 2.3

Chemical Disinfection Techniques

Serial No.	Disinfection Agent	Concentration of Disinfectant	% Inactivation	Time Required	Type of Organism	Reference
1	Free chlorine	0.046 to 0.055 mg/l	100%	1.0 min	*E. coli, S. typhi*	Butterfield et al. (1943)
2	Free chlorine	1.4 mg/l	99.6%	5 min	Coxsackievirus A2	Clarke et al. (1964)
3	Free chlorine	0.10 mg/l	99%	10 min	Poliovirus 1	Weidenkopf (1958)
4	Hypochlorous acid	1.0 mg/l	99%	<10 s	*E. coli*	Scarpino et al. (1972)
5	Hypochlorite ion	1.0 mg/l	99%	50 s	*E. coli*	Scarpino et al. (1972)
6	Monochloramine (NH$_2$Cl)	0.6 mg/l	100%	60 min	*Escherichia coli, Enterobacter aerogenes, Pseudomonas aeruginosa, Salmonella typhi,* and *Shigella dysenteriae*	Butterfield (1948), Butterfield et al. (1943)
7	Monochloramine	1.0 ppm	99%	20 min	*E. coli*	Siders et al. (1973)
8	Dichloramine (NHCl$_2$)	1.2 mg/l	99.999%	10 min	Enteric bacteria	Chang (1971)
9	Ozone	0.5 mg/l	7.5 logs	15 s	*E. coli, Staphylococcus aureus, Salmonella typhimurium, Shigella flexneri, Psuedomonus fluorescens,* and *Vibrio cholerae*	Burleson et al. (1975)

	Disinfectant	Dose	Percentage	Time	Organism	Reference
10	Ozone	0.3 mg/l	99%	8 s	Poliovirus 1	Katzenelson et al. (1974)
11	Ozone	0.5 mg/l of ozone	>99.9%	15 s	Vesicular stomatitis virus, encephalomyocarditis virus, GD VII virus	Burleson et al. (1975)
12	Ozone	0.3 mg/l	98 to >99%	5 min	*Entamoeba histolytica* cysts	Newton and Jones (1949b)
13	Chlorine dioxide	0.25 mg/l	99%	15 s	*E. coli*	Bernarde et al. (1965)
14	Iodine	5 to 10 mg/l	100%	—	Bacterial pathogens	Chang and Morris (1953)
15	Iodine	1 mg/l	>99.9%	1 to 5 min	Enteric bacteria	Chambers et al. (1952)
16	Iodine	10 mg/l	0.5 log / 4 logs	1 h / 1 min	Virus / *E. coli*	Kruse et al. (1970)
17	Iodine	2 mg/l	99%	10 min	Cysts	Chang and Morris (1953)
18	Hypobromous acid	4 mg/l	99%	10 min	*E. coli*	Kruse et al. (1970)
19	Bromine	2 to 25 mg/l (25°C, pH 3)	99%	—	Spores of *Bacillus metiens* and *B. subtilis*	Wyss and Stockton (1947), Marks and Strandskov (1950)
20	Bromine as hypobromous acid	4 mg/l	3.7 logs	10 min	f2 *E. coli* phage virus	Kruse et al. (1970)
21	Hypobromous acid	0.32 mg/l	99%	1.1 min	*E. coli* phage	Taylor and Johnson (1974)
22	Ferrate	6.0 mg/l / 12 mg/l / 60 mg/l	99%	8.5 min / 5 min / 1.5 min	*Escherichia coli* / *Streptococcus faecali*	Gilbert et al. (1976)
23	Ferrate	1.2 mg/l	99%	4 min	RNA bacterial f2 virus	Waite (1978a, 1978b)

(Continued)

TABLE 2.3 (CONTINUED)

Chemical Disinfection Techniques

Serial No.	Disinfection Agent	Concentration of Disinfectant	% Inactivation	Time Required	Type of Organism	Reference
24	Calcium hydroxide	pH 10.5	95%	8 h	*E. coli*	Riehl et al. (1952)
25	Sodium chloride	100 mg/l	99.83% at pH 12.5	5 min	Poliovirus 1	Sproul et al. (1978)
26	Hydrogen peroxide	25,000 mg/l	*S. aureus* by 6 logs Spores by 99%	1 min <1 to 17 min	*S. aureus* and spores of several species of *Bacillus* and *Clostridium*	Toledo et al. (1973)
27	Hydrogen peroxide	3000 mg/l	99%	6 h	Poliovirus	Lund (1963b)
28	Hydrogen peroxide	15,000 mg/l 30,000 mg/l	99%	24 min 4 min	Rhinovirus (types 1A, 1B, and 7)	Mental and Schmidt (1973)
29	Potassium permanganate	1 to 16 mg/l	99%	4 to 120 min	*E. coli*	Cleasby et al. (1964)
30	Silver	100 mg/l	99%	3 to 4 h	(Disinfected water)	Just and Szniolis (1936)
31	Silver	90 mg/l 22.5 mg/l	99%	30 min 60 min	*Vibrio comma*	Yasinskii and Kuznetsova (1973)
32	Silver nitrate	150 mg/l	Moderate	1 to 6 min	Cyst	Chang and Baxter (1955)

described in the section of breakpoint chlorination. This could even result in the removal of coliform bacteria, viruses, and *Giardia* from the water, along with organic matter and heavy metals, if appropriate coagulating or flocculating agents are used. Some of the advantages of coagulation/flocculation can be obtained by allowing the particles to settle out of the water with time (sedimentation), but it will take a while (1 to 2 or several hours) for them to do so. Adding coagulation chemicals such as alum will increase the rate at which the suspended particles settle out by combining many smaller particles into larger flocs, which will settle out faster. The usual dose of alum is 10 to 30 mg/l of water. This dose must be rapidly mixed with the water, followed by gentle agitation for 5 min to encourage the particles to form flocs. After this, at least 30 min of settling time is needed for the flocs to settle to the bottom, and then the clear water above the flocs may be decanted off. Most of the flocculation agent is removed with the floc; nevertheless, some question the safety of using alum due to the toxicity of the aluminium in it. High aluminium toxicity is reported to cause Alzheimer's disease; however, there is little to no scientific evidence to back this up. Many municipal plants do treat the water with alum, as a pretreatment to disinfection. In bulk water treatment, the alum dose can be varied until the ideal dose is found by a technique known as the jar test (*www.nesc.wvu.edu/ndwc/articles/ot/SP05/TB_jartest.pdf*). The needed dose varies with the pH of the water and the size of the already existing particles. Increased turbidity makes the flocs easier to produce, due to the increased number of collisions between particles during gentle agitation, and promote flocculation.

2.7 Disinfection By-Products (DBPs)

Disinfection by-products (DBPs) form when chemical disinfectants react with the organic material (from the decomposition of leaves and other vegetation and microbial cell debris) naturally found in drinking water sources. As discussed extensively in the preceding sections, the use of chlorine to disinfect drinking water has been hailed as one of the major public health breakthroughs in the twentieth century, resulting in a large decrease in mortality from waterborne infectious disease. However, in 1976 the National Cancer Institute (NCI) published data showing that chloroform, a chlorination by-product, caused cancer in rodents (NCI, 1976). There is now quantitative evidence that disinfection, thought pivotal in fighting infectious disease, may also result in cancer and other health risks for humans through its by-product formation.

Two classes of DBPs usually measured in drinking water are trihalomethanes and haloacetic acids (HAAs). THMs, particularly chloroform, have commonly been found in U.S. drinking water supplies. Other by-products also exist but are not regularly measured. Increasing health concerns have triggered widespread research only in the area of DBPs globally, especially in the last decade. It has been estimated that much of the mutagenicity of the DBPs is related to the compound 3-chloro-4-(dichloromethyl)-5-hydroxy-2(5H)-furanone (MX) (Kargalioglu et al., 2002). Tap water monitoring studies in Massachusetts suggest that levels of MX may be considerably higher than previously reported in the United States or Europe (up to 80 ng/l) (Wright et al., 2002). Recent studies indicate that the type of DBPs formed depends on the chemical disinfection employed. Generally, trihalomethanes, haloacetic acids, halonitromethanes (HNMs), haloacetonitriles (HANs), haloaldehydes (HAs), haloketones (HKs), and iodo-THMs (i-THMs) are reported in water treated with chlorine and chloramines. Compared to chlorine, monochloramination generally resulted in lower concentrations of DBPs with the exception of 1,1-dichloropropanone. Another interesting finding was that higher bromide levels lead to higher concentrations of brominated DBPs (Bougeard et al., 2010).

Difficulties arise in balancing microbial risks from contaminated drinking water with the chemical risks posed by DBPs. Many DBPs have been shown to cause cancer and reproductive and developmental effects in animal studies; however, animal toxicity studies generally focus on one chemical by-product delivered in high doses. Epidemiological studies, too, have suggested a weak association between DBPs and cancer and adverse reproductive effects in humans. Such studies offer a crude measurement of what is likely to be a small yet prevalent risk. More than 200 million people in the United States drink chemically disinfected water, so even a relatively small risk may be significant (USEPA, 2001a,b).

2.7.1 Factors Affecting DBP Production

Various characteristics of local water sources and treatment methods contribute to the production of DBPs. The amount of organic precursors in the source water (i.e., the water coming into the treatment plant from lakes, rivers, or wells), water temperature and pH, the amount and type of chemical disinfectants, and the stage in the process at which the disinfection occurs all affect DBP levels (USEPA, 1999). When chlorine has more time to react with the organic matter in the water, levels of THMs and other DBPs increase. For this reason, homes at the outskirts of the distribution system tend to have higher levels of THMs than homes closer to the treatment plant due to the remaining residual chlorine (Steenland and Savitz, 1997). THM levels also vary by season because of the variation in the amount of organic matter in the surface water. Water systems supplied by groundwater will usually have low levels of THMs and other DBPs, because the groundwater contains much less organic matter. Several strategies can help reduce the production of DBPs in drinking water;

for example, pretreatment with activated carbon helps to remove organic compounds, and the use of alternative primary or secondary disinfectants can reduce the by-products. As water purveyors turned toward methods of disinfection other than chlorination (e.g., ozonation, chloramination, chlorine dioxide, UV), more data were needed on the production and health effects of their by-products (Boorman et al., 1999). Richardson in his very recent publication has presented a comprehensive list and discusses a plethora of by-products for major chemical disinfectants, like chlorine, chloramines, ozone, and chorine dioxide, and their combinations. Apart from the usual DBPs known, unregulated DBPs, including halonitromethanes, iodo-acids, haloacids, iodo-trihalomethanes, halomethanes, MX (3-chloro-4-(dichloromethyl)-5-hydroxy-2(5H)-furanone) and brominated MX compounds (BMXs), halonitriles, haloamides, haloaldehydes, haloketones, oxyhalides (including chlorate and iodate), N-nitrosodimethylamine (NDMA) and other nitrosamines, nonhalogenated aldehydes, ketones, and carboxylic acids are discussed in detail. An interesting finding was that other routes of DBP's exposure beyond ingestion (including bathing, showering, and swimming) was also found to be harmful (Richardson, 2011).

In addition to the organic precursors present in water, algae have interestingly been found to contribute to DBPs. It is reported that algae and their extracellular organic matter can be precursors for DBPs. A study on *Scenedesmus quadricauda*, a green alga, and its effect on disinfection by-product formation after ozone disinfection revealed that there was a 10% increase in the THM formation potential. This was attributed to an increase in the dissolved organic carbon content by ozonation of the algal suspension, thereby increasing the concentration of precursors (Plummer and Edzwald, 1998).

2.7.2 Adverse Health Effects of DBPs in Drinking Water

In laboratory studies using rats and mice, THMs have been linked to cancers of the colon and kidney, and haloacetic acids have been linked to liver tumors (Boorman et al., 1999). The risk of cancer from DBPs in humans, especially cancers of the colon, rectum, and bladder, has been documented by several epidemiological studies (Cantor et al., 1998; Doyle et al., 1997; Hildesheim et al., 1998; King and Marrett, 1996). In addition, DBPs are potentially harmful to foetuses. Italian neonates were more likely to have a shorter body length and cranial circumference if their mothers had consumed chlorinated water during pregnancy (Kanitz et al., 1996). A study in Iowa indicated associated intrauterine growth retardation with chloroform levels in drinking water consumed by pregnant women (Kramer et al., 1992). Studies have linked THMs in chlorinated drinking water with an increased frequency of stillbirths (King et al., 2000, reviewed in Bove et al., 2002). One of the best-conducted studies on the reproductive effects of DBP exposure found a strong association between spontaneous abortions and THMs, especially bromodi-chloromethane (Waller et al., 1998). THMs in drinking water have also been linked to birth defects such as neural tube defects (Bove et al., 2002; Klotz and Pyrch, 1999). A review of the epidemiological literature on DBPs and adverse pregnancy outcomes showed the strongest evidence of association with small amounts of DBPs, for gestational age at birth, neural tube defects, and spontaneous abortions (Bove et al., 2002). Associations between DBPs and oral cleft defects and urinary tract defects have also been seen (Bove et al., 2002). DBPs have been shown to be absorbed through dermal and inhala-tion exposure as well; studies indicate that the act of showering increases the levels of DBPs in the bloodstream, consistent with the type of DBPs present in the tap water (Miles et al., 2002). In a recent study, an analysis of the geno-toxic potential of two halogenated acetaldehydes (HAs), namely, tribromoac-etaldehyde (TBA) and chloral hydrate (CH), was conducted in human cells (TK6 cultured cells and peripheral blood lymphocytes). The results showed that both compounds were clearly genotoxic, inducing high levels of DNA breaks, TBA being more effective than CH. Both HAs produced high levels of oxidised bases, and the induced DNA damage was rapidly repaired over time. This is of significance in terms of risk assessment of DBP exposure (Liviac, Creus, and Marcos, 2010). In spite of several epidemiologic studies conducted to ascertain the toxicity of DBPs, no individual DBP can account for the relative risk estimates reported in the positive epidemiologic studies, leading to the hypothesis that these outcomes could result from the toxicity of DBP mixtures. An extensive study in this arena is the need of the hour. A paper published in July 2011 reports a mixture risk assessment for DBP developmental effects. Twenty-four developmentally toxic DBPs were iden-tified, and four adverse developmental outcomes associated with human DBP exposures: spontaneous abortion, cardiovascular defects, neural tube defects, and low-birth-weight infancy were reported. Spontaneous abortion was attributed to the hormonal disruption of pregnancy (Coleman et al., 2011).

Thus, health concerns associated with DBPs appear to be an area that will continue to highlight the research platform of disinfection of water.

2.7.3 Regulation of DBPs in Drinking Water

In order to control DBP concentrations, methods are developed to reduce concentrations of organic materials, such as DBP precursors and DBPs them-selves in water (USEPA, 1999). The USEPA has encouraged these methods and has implemented several rules designed to protect against microbial contaminants while minimising levels of DBPs in drinking water. In its first regulation recommendation regarding DBPs in 1979, the USEPA set a maxi-mum contaminant level for total THMs as an annual average of 0.10 mg/l (USEPA, 2001b). In 1998, the USEPA strengthened the existing rules with its

Stage 1 Disinfectants and Disinfection Byproducts Rule. This rule reduced the maximum contaminant level for total THMs (comprising four contaminants) to 0.080 mg/l (USEPA, 2001b). The rule also sets maximum contaminant levels for five haloacetic acids at 0.060 mg/l, and also sets maximum contaminant levels for chlorite and bromate. USEPA estimates that up to 140 million people will receive more protection because of this rule, and that the levels of THMs will be reduced by 24% nationwide (USEPA, 2001b). Large surface water systems abided by this rule by January 2002, while the compliance deadline for groundwater systems and small surface water systems was met in January 2004. More stringent limitations are also under consideration. The Interim Enhanced Surface Water Treatment Rule requires methods and utilities to identify sources of contamination through assessing source waters and profiling their disinfection by-products. These assessments look at the factors, such as the organic matter in the source water, that influence the amounts and types of DBPs in each drinking water system.

2.7.4 Examples from Literature

Literature reveals a number of case studies on disinfection by-products resulting from chemical disinfections of water. Some of them are listed in Table 2.4.

2.7.5 Emerging Contaminants in Disinfection By-Products

As water disinfection continues to provide safe water to humankind, it also results in the generation of new disinfection by-products, which may also pose a serious health hazard subsequently. In view of this, the USEPA publishes a Contaminant Candidate List (CCL) every 5 years by collection of relevant data from water disinfection units and determines if further analysis and research is warranted. The first CCL was published in 1998, followed by the second CCL in 2003. Currently the third list published in 2008 is available (www.epa.gov). Some of the emerging disinfection by-products, as listed in the current CCL 3, are discussed in this section. The nitrogenous disinfection by-products (N-DBPs) are one of the most potentially lethal contaminants, which include all the nitrogen-containing DBPs. Apart from this, the brominated and iodinated DBPs are also equally harmful. The brominated and iodinated DBPs are formed due to the reaction of the disinfectant with bromide and iodide that may be present in the water being treated (Richardson et al., 2007). The CCL 3 also includes five nitrosamines as emerging contaminants (www.epa.gov/ safewater/ccl/ccl3). An interesting study conducted by Krasner et al. (2006) revealed that haloactealdehydes were the third main class of DBPs formed apart from THMs and HAAs. Interestingly, alternative disinfectants were found to produce fewer amounts of regulated DBPs, such as THMs and HAAs, but they resulted in the formation of other emerging and priority DBPs. This is concerning, as solutions to one kind of a problem result in new issues that

TABLE 2.4

Common Disinfection By-Products Reported in Literature

Disinfecting Agent/ Method of Treatment	By-Product	Reference
Chlorine (0.05 to 1.2 mg/l)	Chloroforms and other trihalomethanes (THMs)	Rook (1974)
	The dihalonitriles (DHANs)	Peters et al. (1990)
	Haloacetic acids	Peters et al. (1991)
	Chloroacetic acid and chloral hydrate	Simpson and Hayes (1998)
Chlorine dioxide (0.2 to 0.5 mg/l)	The inorganic species chlorite and chlorate have been identified as significant by-products	Trussell (1993), Singer (1993)
Chloramines (0.6 to 1.2 mg/l)	Chloropicrin, cyanogen chloride, 1,1-dichloropropanone, and chloramines	Trussell, Zhang, and Raman (1993)
Bromine (2 to 25 mg/l)	Brominated compounds	Ali and Riley (1990)
Ozone (0.3 to 0.5 mg/l)	Aldehydes and glyoxals, bromoform, dibromoacetic acid, cyanogen bromide, and bromate	Cumming and Jolley (1993)

need to be tackled. Plewa et al. (2004) reported for the first time haloamides in chloraminated water. These compounds are reported to be cytotoxic and genotoxic to mammalian cells. N,N-dimethyl-N-nitrosoamine (NDMA), which is potentially carcinogenic and occurs in chlorinated or chloraminated water, is currently not regulated in the United States. However, it has been included in the second Unregulated Contaminants Monitoring Rule (UCMR-2) (www.epa.gov/safewater/ucmr/ucmr2/basicinformation.html).

Human exposure studies are no longer restricted only to potential hazards of ingesting DBP containing water, but are expanding to other routes, such as inhalation and dermal exposure. Researchers have found that exposure to skin by bathing and showering can offer greater exposure to particular DBPs than drinking water (Gordon et al., 2006). Currently, the basic definition of DBPs appears to be changing with the identification of new emerging contaminants. Conventionally a DBP is a compound that is formed in water due to the reaction of natural organic matter (NOM) with the disinfectant. Recently, several pollutant materials are also found to react with the disinfectants, introducing new DBPs in addition to the conventional earlier ones. These are referred to as DBPs of pollutants. Several compounds, such as pharmaceuticals, hormones, pesticides, dyes, surfactants, personal care products, and so on, appear in water, and the fate of these compounds in drinking water is a remarkable area of research in itself. Most of these compounds typically contain an aromatic ring that can react with the chemical disinfectant, such as chlorine and ozone, leading to the formation of the contaminant DBPs. Chlorination of water containing acetaminophen, which is commonly known as paracetamol, resulted in the formation of 1,4-benzoquinone, N-acetyl-p-benzoquinoneimine, chloro-4-acetamidophenol, and dischloro-4-acetamidophenol (Bedner and MacCrehan, 2006). Other reports include those of cimetidine, an antacid, which was found to react with chlorine generating by-products such as cimetidine sulphoxide, 4-hydroxymethyl-5-methyl-1H-imadazole (Buth, Arnold, and McNeil, 2007), and triclosan, an antimicrobial compound routinely used in hand soaps, which reacted with chloramines generating chloroform and other chlorinated phenols (Greyshock and Vikesland, 2006). Parabens, yet another class of antimicrobial compounds employed as preservatives in shampoos, cosmetics, and toothpaste, reacted with chlorine yielding bromo and bromochloroparabens (Canosa et al., 2006). Pesticides also find a way into drinking water systems and are likely to react with the disinfectant used to treat such water systems, leading to reactions involving the pesticide and the chemical disinfectant. This then generates pesticide DBPs, such as those formed by the reaction of an organophosphate pesticide with chlorine, resulting in DBPs such as chlorpyrifos, which in turn oxidises to an oxon reaction product under drinking water treatment conditions normally encountered. The latter product is reported to be much more toxic than the corresponding parent compound (Duirk and Collette, 2006). Dye DBPs (Carneiro et al., 2006; Oliveira et al., 2006) and diesel DBPs (Negreira et al., 2008) are another class of disinfection by-products that have triggered some concern recently owing to their potential mutagenicity. Lastly, it is amazing and alarming to note that UV filters can also react with the residual chlorine present in tap water to generate brominated and chlorinated disinfection by-products (Lebedev, 2007). This really makes one muse over the current scenario of drinking water disinfection and leaves many questions unanswered.

2.7.6 Controlling Disinfection By-Products

Alternatives to chlorination have been suggested, but all alternative chemical methods also form by-products. In addition, because alternative disinfectants cannot provide the residual protection of chlorine-based disinfectants, they must be used in combination with chlorine or chloramines to provide a complete and lasting disinfection treatment. Disinfection by-products can be reduced by removing DBP precursors and protecting source water, where possible, from the entry of DBP precursors, as discussed before. Removing organic precursors through enhanced coagulation and changing the point of chlorination to a later stage in the treatment process are examples of some of the measures that can help control by-product formation. Treatment techniques are available that provide water suppliers the opportunity to maximise potable water safety and quality while minimising the risk of DBP formation. Generally, the best approach to reduce DBP formation is to remove natural organic matter precursors prior to disinfection. USEPA has published a guidelines document for water system operators entitled *Controlling Disinfection Byproducts and Microbial Contaminants in Drinking Water* (USEPA, 2001a,b). The USEPA guidance discusses three processes to effectively remove natural organic matter prior to disinfection:

- *Coagulation, flocculation, and clarification*: Most treatment plants optimise their coagulation process for turbidity (particle) removal. However, coagulation processes can also be optimised for natural organic matter removal with higher doses of inorganic coagulants (such as alum or iron salts), or for organic polymeric flocculants and optimisation of the pH necessary for successful coagulation and flocculation.

- *Adsorption*: Activated carbon can be used to adsorb soluble organics that react with disinfectants to form by-products, prior to the addition of the disinfectants.

- *Membrane technology*: Membranes, used historically to desalinate brackish waters, have also demonstrated excellent removal capabilities of natural organic matter. Membrane processes use hydraulic pressure to force water through a semipermeable membrane that rejects most contaminants, depending on their molecular weights. Variations of this technology include reverse osmosis (RO), nanofiltration (low-pressure RO), and microfiltration (comparable to conventional sand filtration), depending on the pore size of the membranes, which decide the molecular weight cutoff.

Other conventional methods of reducing DBP formation include changing the point of chlorination and using chloramines for residual disinfection. The USEPA predicts that most water systems will be able to achieve compliance with new DBP regulations through the use of one or more of these relatively low-cost modifications in the conventional treatment methods (USEPA, 1998).

Increasing concerns of DBPs in drinking water warrant not only alternative techniques for their control, but also enhanced research in the field of improved efficiency of the existing chemical disinfectant methods. Recently, the formation of nitrogenous disinfection by-products and carbonaceous disinfection by-products (C-DBPs) was investigated upon chlorination of water samples. It was observed that the coagulation-DAF-filtration process resulted in higher removal of algae, dissolved organic nitrogen (DON), and dissolved organic carbon (DOC) than coagulation-IPS-filtration (Hai-Chu et al., 2011).

Thus, in 1974, scientists discovered that during the water treatment process, chlorine reacts with the organic matter in raw water to form DBPs. Other disinfectants also form DBPs. Concerns that the presence of these compounds in drinking water may present potential health risks led the USEPA to propose regulations to control DBPs in 1974. Nevertheless, 25 years of research have failed to establish a direct link between the trace amounts of chlorinated DBPs present in the tap water and any additional cancer risk in humans. In 1990, the International Agency for Research on Cancer evaluated the body of research concerning the potential health effects of chlorinated drinking water and concluded that it is "not classifiable as to its carcinogenicity to humans" (International Agency for Research on Cancer, World Health Organisation, 1991). Furthermore, the World Health Organisation noted that "the risks to health from disinfection by-products are extremely small in comparison with the risks associated with inadequate disinfection, and it is important that disinfection should not be compromised in attempting to control such by-products" (World Health Organisation, 1993). It has been aptly pointed out that "the risk of death from known pathogens in untreated surface water appears to be at least 100 to 1000 times greater than the risk of cancer from known DBPs in chlorinated drinking water" (Regli et al., 1993).

2.8 Conclusion

Protecting public health involves establishing priorities. Comparative risk assessment suggests that the prevention of observed waterborne diseases through the control of microbiological contaminants should take precedence over eliminating the hypothetical risks posed by disinfection by-products. Future drinking water regulations should rely on cost-benefit analysis to determine where the money spent for water treatment will yield the most public health benefits. Humans have a right to expect a safe drinking water supply, but achieving drinking water safety is not simple. Aquatic pathogens, toxic chemicals, heavy metals, and pesticides are some of the threats to our drinking water. As the list of chemical and biological risks expands, so do public concerns. Efforts to reduce one risk in drinking water may introduce either intentionally or unintentionally a different risk to the population using the treated water supply. People have come to fear that chlorine in their drinking water may cause cancer. But chemical disinfectants such as chlorine are used to combat serious waterborne microbial diseases. If we stopped adding these chemicals to drinking water in an effort to reduce the risk of long-term cancer, would we thereby increase the risk of waterborne microbial disease? Would we just trade one form of risk for another? This dilemma is at the core of modern drinking water treatment policy.

Nothing known to science, including the contents of drinking water, is 100% safe. Not only is water an essential element of life, it is for many people an involuntary risk. Much of the population receives water from large community systems over which it has limited control. There are two different involuntary risks posed by modern drinking water: chemical versus microbial contamination. Concern over the potential carcinogenicity of chlorine and its by-products has pushed society to explore other disinfection alternatives. But these options also resulted in their disinfection by-products, thus proving that there can never be a perfect chemical disinfectant. Waterborne enteric pathogens are still responsible for much greater levels of illness than are chemicals or other contaminants in drinking water (Gerba and Haas, 1988). On the other hand, even a small additional risk of cancer or other adverse health effects from disinfecting agents could eventually account for a significant amount of latent illness (NAS, 1987). Given the risks of microbes and chemicals in drinking water, and the disinfection technologies available at present, there is no simple solution. We need to weigh all of the risks of drinking water in a thoughtful, sensible manner and search for solutions that reduce the overall risks. Physical techniques and hybrid methods involving lower use of disinfecting chemicals could be a potential method for water disinfection, and these have been dealt with in detail in the succeeding chapters.

Questions

1. Define *chemical disinfection* and explain its theory.
2. Explain the need and role of mathematical models for chemical disinfection.
3. Discuss the Chick–Watson model for chemical disinfection.
4. Describe briefly the history of chlorination, highlighting the typical milestones.
5. Explain breakpoint chlorination and its significance.
6. How does chlorine inactivate microorganisms? Discuss the possible mechanisms proposed in the literature.
7. What are the factors that affect chlorine disinfection?
8. Describe various methods of chlorine treatment for potable water.
9. What are the typical merits and demerits of various chlorine-based disinfection methods for potable water? Among the existing methods, which do you think is a good option?
10. Explain the need for newer chemical disinfection technologies for water disinfection.

11. How does iodine kill pathogens? Mention the typical concentration levels used and discuss its merits and limitations as a disinfectant.

12. Explain the role of silver in water disinfection emphasising the recent trends and developments in its application as a disinfectant.

13. Describe potassium permanganate as a chemical disinfectant.

14. Write short notes on disinfection by:

    a. Hydrogen peroxide

    b. Bromine

    c. Ferrate

15. Explain ozone generation and its role as a chemical disinfectant for water disinfection.

16. Answer the following based on the literature data presented in Table 2.3.

    a. Enlist all the chemical disinfection methods used for inactivation of *E. coli*. Comment on the best method for the same and justify your answer.

    b. Which is the best method for inactivation of viruses and why?

    c. Can cysts be inactivated using chemical methods? Which chemical method would you recommend?

    d. If the water disinfection criterion is 99% microbial inactivation, which method would you recommend based on the concentration of chemicals required and the time of treatment? Justify your answer.

17. Does the process of water disinfection require pretreatment? If yes, discuss the typical water pretreatment required before chemical disinfection.

18. Discuss various DBPs formed during chemical disinfection of water.

19. Explain the factors that affect the disinfection by-product formation.

20. What are the potential risks and possible remedies for disinfection by-products in potable water?

References

Ali, M.Y.A., and Riley, J.P. (1990). Distribution of halomethanes in potable waters of Kuwait. *Water Res.* 24(4):533–538.

Allen, T.L., and Keefer, I.L.M. (1955). The formation of hypoiodous acid and hydrated iodine cation by the hydrolysis of iodine. *J. Am. Chem. Soc.* 77:2957–2960.

Alowitz, M., and Scherer, M. (2002). Kinetics of nitrate, nitrite, and Cr(VI) reduction by iron metal. *Environ. Sci. Technol.* 36:299.

American Water Works Association. (1971). *Water quality and treatment: A handbook of public water supplies*, 3rd ed. New York: McGraw Hill.

American Water Works Association. (2000). Water Quality Division Disinfection Systems Survey Committee report. *J. Am. Water Works Assoc.* 9:24–43.

Audette, R.J., Quail, J.W., and Smith, P.J. (1971). Ferrate VI ion, a novel oxidizing agent. *Tetrahedron Lett.* 3:279–281.

Bao, Q., Zhang, D., and Qi, P. (2011). Synthesis and characterization of silver nanoparticle and graphene oxide nanosheet composites as a bactericidal agent for water disinfection. *J. Colloid Interface Sci.* 360(2):463–470.

Becarud, N. (1966). *Analytical study of ferrates*. Comm. Energy at France Rapt. 2895.

Bedner, M., and MacCrehan, W.A. (2006). Transformation of acetaminophen by chlorination produces the toxicants 1,4-benzoquinone and N-acetyl-p-benzoquinone imine. *Environ. Sci. Technol.* 40(2):516–522.

Bell, R.P., and Gelles, E. (1951). The halogen cations in aqueous solution. *J. Chem. Soc.* III:2736–2740.

Benarde, M.A., Snow, W.B., Olivieri, V.P., and Davidson, B. (1967). Kinetics and mechanism of bacterial disinfection by chlorine dioxide. *Appl. Microbiol.* 15:257–265.

Berg, G., Chang, S.L., and Harris, E.K. (1964). Devitalization of microorganisms by iodine. 1. Dynamics of the devitalization of enteroviruses by elemental iodine. *Virology* 22:46.

Bocharov, D.A., and Kulikovskii, A.V. (1971). Structural and biochemical changes of bacteria after the action of some chlorine-containing preparations. Report 1. *Tr. Vses. Nauchno-Issled. Inst. Vet. Sanit.* 38:165–170.

Boorman, G.A., Dellarco, V., Dunnick, J.K., Chapin, R.E., Hunter, S., and Hauchman, F. (1999). Drinking water disinfection byproducts: Review and approach to toxicity evaluation. *Environ. Health Perspect.* 107(Suppl 1):207–217.

Bougeard, C.M.M., Goslan, E.H., Jefferson, B., and Parsons, S.A. (2010). Comparison of the disinfection by-product formation potential of treated waters exposed to chlorine and monochloramine. *Water Res.* 44(3):729–740.

Bove, F., Shim, Y., and Zeitz, P. (2002). Drinking water contaminants and adverse pregnancy outcomes: A review. *Environ. Health Perspect.* 110(Suppl. 1):61–74.

Brammer, K.W. (1963). Chemical modification of viral ribonucleic acid. II. Brominstion and iodination. *Biochim. Biophys. Acta* 72:217–229.

Bruzzoniti, M.C., Kobylinska, D.K., Franko, M., and Sarzanini, C. (2010). Flow injection method for the determination of silver concentration in drinking water for spacecrafts. *Anal. Chim. Acta* 665(1):69–73.

Burleson, G.R., Murray, T.M., and Pollard, M. (1975). Inactivation of viruses and bacteria by ozone, with and without sonication. *Appl. Microbiol.* 29:340–344.

Butterfield, C.T. (1948). Bactericidal properties of chloramines and free chlorine in water. *Public Health Rep.* 63:934–940.

Butterfield, C.T., Wattie, E., Megregien, S., and Chambers, C.W. (1943). Influence of pH and temperature on the survival of coliforms and enteric pathogens when exposed to free chlorine. *Public Health Rep.* 58:1837–1866.

Canosa, P., Rodriguez, I., Rubi, E., Negreira, N., and Cela, R. (2006). Formation of halogenated by-products of parabens in chlorinated water. *Anal. Chim. Acta* 575:106–113.

Cantor, K.P., Lynch, C.F., Hildesheim, M.E., Dosemeci, M., Lubin, J., and Alavanja, M. (1998). Drinking water source and chlorination byproducts. I. Risk of bladder cancer. *Epidemiology* 9:21–28.

Cao, J., Elliott, D.W., and Zhang, W. (2005). Perchlorate reduction by nanoscale iron particles. *J. Nanopart. Res.* 7:499.

Carneiro, P.A., Rech, C.M., Zanoni, M.V.B., Claxton, D., and Umbuzeiro, G.A. (2006). Mutagenic compounds generated from the chlorination of disperse azo-dyes and their presence in drinking water. *Environ. Sci. Technol.* 40(21):6682–6689.

Chadwick, A.F., and Hoh, G.L.K. (1966). Hydrogen peroxide. In *Kirk-Othmer encyclopedia of chemical technology*, Vol. II, 2nd ed. New York: Wiley Interscience, pp. 319–417.

Chambers, C.W., Kabler, P.W., Malaney, G., and Bryant, A. (1952). Iodine as a bactericide. *Soap Sanit. Chem.* 28(10):149–151.

Chang, S.L. (1944). Destruction of microorganisms. *J. Am. Water Works Assoc.* 36:1192–1207.

Chang, S.L. (1970). Modern concepts of disinfection. In *Proceedings of the National Speciality Conference on Disinfection*. New York: American Society of Civil Engineers, pp. 635–681.

Chang, S.L. (1971). Modern concepts of disinfection. *J. Sanit. Eng. Div. AZEL Soc. Civ. Eng.* 97:680–707.

Chang, S.L., and Baxter, M. (1955). Studies on destruction of cysts of *Entamoeba histolytica*. I. Establishment of the order of reaction in destruction of cysts of *E. histolytica* by element iodine and silver nitrate. *Am. J. Hyg.* 61:121–132.

Chang, S.L., and Morris, J.C. (1953). Elemental iodine as a disinfectant for drinking water. *Ind. Eng. Chem.* 45:1000–1012.

Chen, J.J., and Yeh, H.H. (2005). The mechanisms of potassium permanganate on algae removal. *Water Res.* 39(18):4420–4428.

Clarke, N.A., Berg, O., Kabler, P.W., and Chang, S.L. (1964). Human enteric viruses in water source, survival and removal and removability. In W.W. Eckenfelder (ed.), *Advances in water pollution research: Proceedings of the 1st International Conference on Water Pollution Research*, London, September 1962. New York: Macmillan, Vol. 2, pp. 523–541.

Cleasby, J.I., Baumann, E.R., and Black, C.D. (1964). Effectiveness of potassium permanganate for disinfection. *J. Am. Water Works Assoc.* 56:466–474.

Coleman, J., Rice, G.E., Wright, J.M., Hunter III, E.S., Teuschler, L.K., Lipscomb, J.C., Hertzberg, R.C., Simmons, J.E., Fransen, M., Osier, M., and Narotsky, M.G. (2011). Identification of developmentally toxic drinking water disinfection byproducts and evaluation of data relevant to mode of action. *Toxicol. Appl. Pharmacol.* 254(2):100–126.

Coleman, P.K. (2011). Abortion and mental health: Quantitative synthesis and analysis. *BJP* 199:180–186.

Connell, G.F. (1996). *The chlorination/chloramination handbook*. Denver: American Water Works Association.

Cumming, R.B., and Jolley, R.L. (1993). Occurrence and exposures to disinfectants and disinfection by-products. In *Safety of water disinfection: Balancing chemical and microbial risks*. ILSI, pp. 257–275.

De Gusseme, B., Hennebel, T., Christiaens, E., Saveyn, H., Verbeken, K., Fitts, J.P., Boon, N., and Verstraete, W. (2011). Virus disinfection in water by biogenic silver immobilized in polyvinylidene fluoride membranes. *Water Res.* 45(4):1856–1864.

Dennis, W.H. (1977). The mode of action of chlorine on f2 bacterial virus during disinfection. Sc.D. thesis, School of Hygiene and Public Health, Johns Hopkins University, Baltimore.

Doyle, T.J., Zheng, W., Cerhan, J.R., Hong, C.P., Sellers, T.A., and Kushi, L.H. (1997). The association of drinking water source and chlorination by-products with cancer incidence among postmenopausal women in Iowa: A prospective cohort study. *Am. J. Public Health* 87:1168–1176.

Drinking Water Standards. (2004). http://www.mass.gov/dep/water/drinking/ standards (accessed May 13, 2004).

Dror-Ehre, A., Adin, G., Markovich, H., and Mamane, H. (2010). Control of biofilm formation in water using molecularly capped silver nanoparticles. *Water Res.* 44(8):2601–2609.

Duirk, S.E., and Collette, T.W. (2006). Degradation of chlorpyrifos in aqueous chlorine solutions: Pathways, kinetics, and modeling. *Environ. Sci. Technol.* 40(2):546–551.

Ellis, K.V., Cotton, A.P., and Khowaja, M.A. (1993). Iodine disinfection of poor quality waters. *Water Res.* 27(3):369–375.

Ellis, K.V., and Van Vree, H.B.R.J. (1989). Iodine used as water disinfectant in turbid water. *Water Res.* 23(6):671–676.

Fair, G.M., Morris, J.C., Chang, S.L., Weil, I., and Burden, R.P. (1948). The behavior of chlorine as a water disinfectant. *J. AZXL Water Works Assoc.* 40:1051–1061.

Fetner, R.H. (1962). Chromosome breakage in *Vicia faba* by monochloramine. *Nature* 196:1122–1123.

Fitzgerald, G.P. (1964). Laboratory evaluation of potassium permanganate as an algicide for water reservoirs. *Southwest Water Works. J.* 45(10):16–17.

Fremy, E.G. (1841). Studies of the action of alkaline peroxides on metal oxides [in French]. *C. R. Acad. Sci. Ser. A* 12:23–24.

Gangadharan, D., Harshvardan, K., Gnanasekar, G., Dixit, D., Popat, K.M., and Anand, P.S. (2010). Polymeric microspheres containing silver nanoparticles as a bactericidal agent for water disinfection. *Water Res.* 44(18):5481–5487.

Gasset, A.R., Ramer, R.M., and Katrin, D. (1975). Hydrogen peroxide sterilization of hydrophilic contact lenses. *Arch. Opthalmol.* 93:412–415.

Gerba, C.P., and Haas, C.N. (1988). Assessment of risks associated with enteric viruses in contaminated drinking water. In J.J. Lichtenberg et al. (eds.), Chemical and biological characterization of sludges, sediments, dredge soils and drilling mud. *ASTM STP* 976:489–494.

Gilbert, M.D., Waite, T.D., and Hare, C. (1976). An investigation of the applicability of ferrate ion for disinfection. *J. Am. Water Works Assoc.* 68:495–497.

Gordon, S.M., Brinkman, M.C., Ashley, D.L., Blount, B.C., Lyu, C., Masters, J., and Singer, P.C. (2006). Changes in breath trihalomethane levels resulting from household water-use activities. *Environ. Health Perspect.* 114(4):514–521.

Green, D.E., and Stumpf, P.K. (1946). The mode of action of chlorine. *J. Am. Water Works Assoc.* 38:1301–1305.

Greyshock, A.E., and Vikesland, P.J. (2006). Triclosan reactivity in chlorinated waters. *Environ. Sci. Technol.* 40(8):2615–2622.

Haas, C.N. (1978). Mechanisms of inactivation of new indicators of disinfection efficiency by free available chlorine. Ph.D. thesis, Department of Civil Engineering, University of Illinois at Urbana–Champaign.

Haas, C.N., and Karra, S.B. (1984). Kinetics of microbial inactivation by chlorine. I. Review of results in demand-free systems. *Water Res.* 18(11):1443–1449.

Hai-Chu, W., Gao, N.-Y., Templeton, M.R., and Yin, D.-Q. (2011). Comparison of inclined plate sedimentation and dissolved air flotation for the minimization of subsequent nitrogenous disinfection by-product formation. *Chemosphere* 83(5):647–651.

He, F., and Zhao, D.Y. (2005). Preparation and characterization of a new class of starch-stabilized bimetallic nanoparticles for degradation of chlorinated hydrocarbons in water. *Environ. Sci. Technol.* 39:3314–3326.

Heiner, J.D., Hile, D.C., Demons, S.T., and Wedmore, I.S. (2010). 10% povidone-iodine may be a practical field water disinfectant. *Wilderness Environ. Med.* 21(4):332–336.

Henderson, L.J. (1935). Physician and patient as a social system. *New England J. Med.* 212:819–820.

Hildesheim, M.E., Cantor, K.P., Lynch, C.F., Dosemeci, M., Lubin, J., and Alavanja, M. (1998). Drinking water source and chlorination byproducts. II. Risk of colon and rectal cancer. *Epidemiology* 9:29–35.

Hooper, G. (1987). Chemical disinfection. In Lorch, W. (ed.) *Handbook of water purification*, 2nd ed. Ellis Horwood Series in Water and Wastewater Technology. Chichester: Inglaterra.

Ingols, I.L.S., and Ridenour, G.M. (1948). Chemical properties of chlorine dioxide in water treatment. *J. Am. Water Works Assoc.* 40:1207–1227.

Ingols, R.L., Wyckoff, H.A., Kethley, T.W., Hodgen, H.W., Fincher, E.L., Hildebrand, J.C., and Mandel, J.E. (1953). Bactericidal studies of chlorine. *Md. Eng. Chem.* 45:996–1000.

Jar Test. (2011). http://www.nesc.wvu.edu/ndwc/articles/ot/SP05/TB_jartest.pdf (accessed October 13, 2011).

John, D.E., Haas, C.N., Nwachuku, N., and Gerba, C.P. (2005). Chlorine and ozone disinfection of *Encephalitozoon intestinalis* spores. *Water Res.* 39(11):2369–2375.

Johnson, J.D., and Overby, R. (1971). Bromine and bromamine disinfection chemistry. *J. Sanit. Eng. Div. Am. Soc. Civ. Eng.* 97:617–628.

Just, J., and Szniolis, A. (1936). Germicidal properties of silver in water. *J. Am. Water Works Assoc.* 28:492–506.

Kaminski, J.J., Huycke, M.M., Sell, S.H., Bodor, N., and Higuchi, T. (1976). N-Halo derivatives. V. Comparative antimicrobial activity of soft N-chloramine systems. *J. Pharm. Sci.* 65:1737–1742.

Kanel, S., Manning, B., Charlet, L., and Choi, H. (2005). Removal of arsenic-(III) from groundwater by nanoscale zero-valent iron. *Environ. Sci. Technol.* 39:1291–1298.

Kanitz, S., Franco, Y., Patrone, V., Caltabellotta, M., Raffo, E., and Riggi, C. (1996). Association between drinking water disinfection and somatic parameters at birth. *Environ. Health Perspect.* 104:516–520.

Kanyaev, N., and Shilov, E.A. (1940). Constants of some equilibrium reactions of hypobromous acid. *Tr. Ivanov. Khim. Tekhnol. Inst.* (USSR) 3:69–73.

Kargalioglu, Y., McMillan, B.J., Minear, R.A., and Plewa M.J. (2002). An analysis of the cytotoxicity and mutagenicity of drinking water disinfection by-products in *Salmonella typhimurium. Teratog. Carcinog. Mutagen* 22:113–128.

Katzenelson, E., Klettel, B., and Shuval, H.I. (1974). Inactivation kinetics of viruses and bacteria in water by use of ozone. *J. Am. Water Works Assoc.* 66:725–729.

Kazama, F. (1995). Viral inactivation by potassium ferrate. *Water Sci. Technol.* 31:165–168.

Kemp, H.T., Fuller, R.G., and Davidson, R.S. (1966). Potassium permanganate as an algicide. *J. Am. Water Works Assoc.* 58:255–263.

King, W.D., Dodds, L., and Allen, A.C. (2000). Relation between stillbirths and specific chlorination by-products in public water supplies. *Environ. Health Perspect.* 108:883–886.

King, W.D., and Marrett, L.D. (1996). Case-control study of bladder cancer and chlorination byproducts in treated water (Ontario, Canada). *Cancer Causes Control* 7:596–604.

Klotz, J.B., and Pyrch, L.A. (1999). Neural tube defects and drinking water disinfection by-products. *Epidemiology* 10:383–390.

Kramer, M.D., Lynch, C.F., Isacson, P., and Hanson, J.W. (1992). The association of waterborne chloroform with intrauterine growth retardation. *Epidemiology* 3:407–413.

Krasner, S.W., Weinberg, H.S., Richardson, S.D., Pastor, S.J., Chinn, R., Sclimenti, M.J., and Kruse, C.W. (1969). Mode of action of halogens on bacteria and viruses and protozoa in water systems. In *Final Report to the Commission on Environmental Hygiene of the Armed Forces Epidemiological Board*. U.S. Army Medical Research and Development Command Contract DA-49-193-MD 2314, pp. 1–89.

Kruse, C.W., Hsu, Y., Griffiths, A.C., and Stringer, R. (1970). Halogen action on bacteria, viruses and protozoa. In *Proceedings of the National Speciality Conference on Disinfection*. New York: American Society of Civil Engineers, pp. 113–136.

Labas, M.D., Zalazar, C.S., Brandi, R.J., and Cassano, A.E. (2008). Reaction kinetics of bacteria disinfection employing hydrogen peroxide. *Biochem. Eng. J.* 38(1):78–87.

Labatuik, C.W., Belosevic, M., and Finch, G.R. (1994). Inactivation of *Giardia muris* using ozone and ozone-hydrogen peroxide. *Ozone Sci. Eng.* 16:67–78.

Langlais, B., Reckhow, D.A., and Brink, D.A. (eds.). (1991). *Ozone in water treatment: Application and engineering*. American Water Works Research Foundation, Lewis Publishers.

Lebedev, A. (2007). Mass spectrometry in the study of mechanisms of aquatic chlorination of organic substrates. *Eur. J. Mass Spectrom.* 13:51–56.

Li, Q., Mahendra, S., Lyon, D.Y., Brunet, L., Liga, M.V., Li, D., and Alvarez, P.J.J. (2008). Antimicrobial nanomaterial for water disinfection and microbial control: Potential applications and implications. *Water Res.* 42:4591–4602.

Li, X.-Q., Elliott, D.W., and Zhang, W.-X. (2006). Zero-valent iron nanoparticles for abatement of environmental pollutants: Materials and engineering aspects. *Critical Rev. Solid State and Mat. Sci.* 31:111–122.

Liebhafsky, H.A. (1934). The equation constant of the bromine hydrolysis and its variations with temperature. *J. Am. Chem. Soc.* 56:1500–1505.

Liviac, D., Creus, A., and Marcos, R. (2010). DNA damage induction by two halogenated acetaldehydes, by-products of water disinfection. *Water Res.* 44:2638–2646.

Lund, E. (1963a). Oxidative inactivation of poliovirus at different temperatures. *Arch Ges. Virusforsch.* 13:375–386.

Lund, E. (1963b). Significance of oxidation in chemical inactivation of poliovirus. *Arch. Ges. Virusforsch.* 12:648–660.

Lund, E. (1966). Oxidative inactivation of adenovirus. *Arch. Ges. Virusforsch.* 19:32–37.

Marks, H.C., and Strandskov, F.B. (1950). Halogens and their mechanism of action. *Ann. N.Y. Acad. Sci.* 53:163–171.

Masschelein, W. (1967). Developments in the chemistry of chlorine dioxide and its applications. *Chim. Ind. Genie Chim.* 97:4941.

Maynard, A.D. (2007). Nanotechnology—Toxicological issues and environmental safety. In *Project on Emerging Nanotechnologies*, 1–14.

Maizel, J.V., Phillips, B.A., and Summers, D.S. (1967). Composition of artificially produced and naturally occurring empty capsids of poliovirus type I. *Virology* 32:692–699.

Mental, R., and Schmidt, J. (1973). Investigations on rhinovirus inactivation by hydrogen peroxide. *Acta Virol.* 17:351–354.

Miles, A.M., Singer, P.H., Ashley, D.L., Lynberg, M.C., Mendola, P., and Langlois, P.H. (2002). Comparison of trihalomethanes in tap water and blood. *Environ. Sci. Technol.* 36:1692–1698.

Mills, J.F. (1975). Interhalogens and halogen mixtures as disinfectants. In J.D. Johnson (ed.), *Disinfection water and wastewater*. Ann Arbor; MI: Ann Arbor Science, Chapter 6.

Mills, J.F. (1969). The control of microorganisms with polyhalide resins. U.S. Patent 3462363.

Momba, M.N.B., Cloete T.E., Venter, S.N., and Kfir, R. (1998). Evaluation of the impact of disinfection processes on the formation of biofilms in potable surface water distribution systems. *Water Sci. Technol.* 38(8/9):283–289.

Morris, J.C. (1970). *Modern chemical methods. I. International courses in hydraulic and sanitary engineering*. Delft, Netherlands: International Institute of Hydraulics and Environmental Engineering.

Morris, J.C. (1978). *Modern chemical methods in water and waste treatment*. Vol. 1. Delft, The Netherlands: International Institute for Hydraulic and Environmental Engineering.

Murmann, R.K. (1974). *The preparations and oxidative properties of ferrate (FeO42–). Studies directed towards its use as a water purifying agent*. NTIS Publication PB Report 238-057. Springfield, VA: National Technical Information Service.

Negreira, N., Canosa, P., Rodriguez, I., Ramil, M., Rubi, E., and Cela, R. (2008). *J. Chromatogr. A* 1178:206–214.

Newton, W.J., and Jones, M.F. (1949a). Effectiveness of silver ions against cysts of *Entamoeba histolytica*. *J. Am. Water Works Assoc.* 41:1027–1034.

Newton, W.L., and Jones, M.F. (1949b). Effect of ozone in water on cysts of *Endamoeba histolytic*. *Am. J. Trop. Med.* 29:669–681.

Nusbaum, I. (1952). Sewage chlorination mechanism. A survey of fundamental factors. *Water Sewage Works* 99:294–297.

Olivieri, V.P., Kruse, C.W., Hsu, Y.C., Griffiths, A.C., and Kawata, K. (1975). The comparative mode of action of chlorine, bromine and iodine on f2 bacterial viruses. In J.D. Johnson (ed.), *Disinfective water and wastewater*. Ann Arbor, MI: Ann Arbor Science Publishers, pp. 145–162.

Page, K., Palgrave, R.G., Parkin, I.P., Wilson, M., Savinand, S.L.P., and Chadwick, A.V. (2006). Titania and silver–titania composite films on glass—Potent antimicrobial coatings. *J. Mat. Chem.* 17(1):95–104.

Peters, R.J.B., De Leer, E.D.W.B., and De Galan, L. (1990). Dihaloacetonitriles in Dutch drinking waters. *Water Res.* 24(6):797–800.

Peters, R.J.B., Erkelens, C., De Leer, E.D.W.B., and De Galan, L. (1991). The analysis of halogenated acetic acids in Dutch drinking water. *Water Res.* 25(4):473–477.

Plewa, M.J., Wagner, E.D., Richardson, S.D., Thruston Jr., A.D., Woo, Y.-T., and McKague, A.B. (2004). Chemical and biological characterization of newly discovered iodoacid drinking water disinfection byproducts. *Environ. Sci. Technol.* 38(18):4713–4722.

Plummer J., and Edzwald, J. (1998). Effect of ozone on disinfection by-product formation of algae. *Water Sci. Technol.* 37:49–55.

Rabea, E.I., Badawy, M.E., Stevens, C.V., Smagghe, G., and Steurbaut, W. (2003). Chitosan as antimicrobial agent: Applications and mode of action. *Biomacromolecules* 4(6):1457–1465.

Rahn, O. (1945). Injury and death of bacteria by chemical agents. *Biodynamica Monograph No. 8.*

Reddy, M.P., Venugopal, A., and Subrahmanyam, M. (2007). Hydroxyapatite-supported Ag–TiO2 as *Escherichia coli* disinfection photocatalyst. *Water Res.* 41:379–386.

Regli, S., Berger, P., Macler, B., and Haas, C. (1993). Proposed decision tree for management of risks in drinking water: Consideration for health and socioeconomic factors. In *Safety of water disinfection: Balancing chemical and microbial risks.* Washington, DC: ILSI, pp. 39–80.

Reynolds, G.W., Hoff, J.T., and Gillham, R.W. (1990). Sampling bias caused by materials used to monitor halocarbons in groundwater. *Environ. Sci. Technol.* 24:135–142.

Richardson, S.D. (2011). Disinfection by-products: Formation and occurrence in drinking water. In *Encylopedia of environmental health.* Burlington, MA: Elsevier Science, pp. 110–136.

Richardson, S.D., Plewa, M.J., Wagner, E.D., Schoeny, R., and DeMarini, D.M. (2007). Sampling bias caused by materials used to monitor halocarbons in groundwater. *Mutat. Res.* 636(1–3):178–242.

Riesser, V.N., Perrich, J.R., Silver, E.B., and McCammon, J.R. (1977). Possible mechanisms of poliovirus inactivation by ozone. In E.G. Fochtman, R.G. Rice, and M.F. Browning (eds.), *Forum on ozone disinfection.* Cleveland, OH: International Ozone Institute, pp. 18–192.

Robinson, P.R., and R. K. Murmann, R.K. (1975). Rapid oxygen exchange between oxohydroxytetrakis(cyano)molybdate(4-) ion and water. *Inorg. Chem.* 14:203.

Ronen, Z., Guerrero, A., and Gross, A. (2010). Grey water disinfection with the environmentally friendly hydrogen peroxide plus (HPP). *Chemosphere* 78(1):61–65.

Rook, J.J. (1974). Formation of haloforms during chlorination of natural waters. *Water Treat. Exam.* 23:234–243.

Rosenkranz, H.S. (1973). Sodium hypochlorite and sodium perborate: Preferential inhibitors of DNA polymerase-deficient bacteria. *Mutat. Res.* 21:171–174.

Ryan, J.N., Harvey, R.W., Metge, D., Elimelech, M., Navigato, T., and Pieper, A.P. (2002). Field and laboratory investigations of inactivation of viruses (PRD1 and MS2) attached to iron oxide-coated quartz sand. *Environ. Sci. Technol.* 36(11):2403–2413.

Scarpino, P.V., Berg, O., Chang, S.L., Dahling, D., and Lucas, M. (1972). A comparative study of the inactivation of viruses in water by chlorine. *Water Res.* 6:95.

Schreyer, J.M., and Ockerman, L.T. (1951). Stability of the ferrate (VI) ion in aqueous solution. *Anal. Chem.* 23:1312–1314.

Schumb, W.C., Satterfield, C.N., and Wentworth, R.L. (1955). *Hydrogen peroxide.* Reinhold: New York.

Scott, D.B. and Lesher, E.C. (1963). Effect of ozone on survival and permeability of *Escherichia coli. J. Bacteriol.* 85:567–576.

Sezgin, M.D., Jenkins, D., and Parker, D.S. (1978). A unified theory of filamentous activated sludge bulking. *J. Water Pollut. Control Fed.* 50:362–381.

Shih, K.L., and Lederberg, J. (1976a). Chloramine mutagenesis in *Bacillus subtitlis. Science* 192:1141–1143.

Shih, K.L., and Lederberg, J. (1976b). Effects of chloramine on *Bacillus subtilis* deoxyribonucleic acid. *J. Bacteriol.* 125:934–945.

Shull, K.E. (1962). Operating experiences at Philadelphia suburban treatment plants. *J. Am. Water Works Assoc.* 54:1232–1240.

Siders, D.L, Scarpino, P.V., Lucas, M., Berg, G., and Qiang, S.L. (1973). Destruction of viruses and bacteria in water by monochloramine. In Abstracts of the *American Society for Microbiology Annual Meeting*, Washington, DC, abstract 7.

Simpson, K.L., and Hayes, K.P. (1998). Drinking water disinfection by-products: An Australian perspective. *Water Res.* 32(5):1522–1528.

Singer, P.C. (1993). Basin Concepts of Disinfection By-Product Formation and Control. *AWWA D/DBP Rule Teleconference: Presentation 1*, pp. 1–20.

Skvortsova, E.I.L., and Lebedeva, N.S. Cited in Bekhtereva, M.N., and Krainova, O.A. (1973). Changes in the activity of enzyme systems in *Bacillus anithracoides* spores during germination and due to the action of calcium hypochlorite. *Mikrobiologiya* 44:791–795.

Smith, D.K. (1967). Disinfection and sterilization of polluted water with ozone. In *Proceedings of 2nd Annual Symposium on Water Research*, McMasters University, Hamilton, Ontario, p. 52.

Sohn, K., Kang, S.W., Ahn, S., Woo, M., and Yang, S.K. (2006). Fe(0) nanoparticles for nitrate reduction: Stability, reactivity, and transformation. *Environ. Sci. Technol.* 40(17):5514–5519.

Spaulding, E.H., Cundy, K.R., and Turner, F.J. (1977). Chemical disinfection of medical and surgical materials. In S.S. Block (ed.), *Disinfection, sterilization and prevention*, 2nd ed. Philadelphia: Lea and Febiger, pp. 654–684.

Spicher, R.G., and Skrinde, R.T. (1963). Potassium permanganate oxidation of organic contaminants in water supplies. *J. Am. Water Works Assoc.* 55:1174–1194.

Sproul, O.J., Emerson, H.A., Howser, H.A., Boyce, D.M., Walsh, D.S., and Buck, C.E. (1978). Effects of particulate matter on virus inactivation by ozone. Presented at the 98th Annual Conference of the American Water Works Association in Atlantic City, NJ.

Steenland, K., and Savitz, D.A. (eds.). (1997). *Topics in environmental epidemiology*. New York: Oxford University Press.

Strong, A.W. (1973). An exploratory work on the oxidation of ammonia by potassium ferrate (VI). M.S. thesis, Department of Chemical Engineering, Ohio State University.

Stumm, W., and Morgan, J.J. (1996). *Aquatic chemistry*, 3rd ed. New York: John Wiley & Sons.

Symons, J.M., Carswell, J.K., Clark, R.M., Dorsey, P., Geidreich, E.E., Heffernarn, W.P., Hoff, I.C., Love, O.T., and Stevens, A. (1977). *Ozone, chlorine dioxide, and chloramines as alternatives to chlorine for disinfection of drinking water: State of the art*. Cincinnati, OH: Water Supply Research Division, USEPA.

Taylor, G.R., and Butler, M. (1982). A comparison of the virucidal properties of chlorine, chlo-dioxide, bromine chloride and iodine. *J. Hyg. Camb.* 89:321–328.

Taylor, D.G., and Johnson, J.D. (1974). A.J. Rubin (ed.), *Chemistry of water supply, treatment and distribution*. Ann Arbor, MI: Ann Arbor Science, pp. 369–408.

Toledo, R.T. (1975). Chemical sterilants for aseptic packaging. *Food Technol.* 29:102, 104, 105, 108, 110, 112.

Toledo, R.T., Escher, F.E., and Ayres, J.C. (1973). Sporicidal properties of hydrogen peroxide against food spoilage organisms. *Appl. Microbiol.* 26:592–597.

Trussell, L.O., Zhang, S., and Raman, I.M. (1993). Desensitization of AMPA receptors upon multiquantal neurotransmitter release. *Neuron IO*. 1185–1196.

USEPA. (1978). *Interim primary drinking water regulations*. Washington, DC: USEPA.

USEPA. (1998, December 16), National primary drinking water regulations: Disinfectants and disinfection by-products: Final Rule, *Fed. Registr.* 63:69389.

USEPA. (1999). *Stage 2 microbial/disinfection byproducts health effects workshop.* Washington, DC: USEPA. Available at http://www.epa.gov/safewater/mdbp/st2feb99.html (accessed April 20, 2002).

USEPA. (2001a). *Drinking water priority rulemaking: Microbial and disinfection byproduct rules.* EPA 816-F-01-012. Washington, DC: USEPA.

USEPA. (2001b). *Stage 1 disinfectants and disinfection byproducts rule.* EPA 816-F-01-014. Washington, DC: USEPA.

USEPA. (2007). In Science Policy Council (ed.), *US Environmental Protection Agency nanotechnology white paper.* EPA 100/B-07/001. Washington, DC: USEPA.

Van Elsen, A., and Boeye, A. (1966). Disruption of type I poliovirus under alkaline conditions: Role of pH, temperature and sodium dodecyl sulphate (SDS). *Virology* 28:481–483.

Venkobachar, C. (1975). Biochemical model for chlorine disinfection. Ph.D. thesis.

Venkobachar, C., Iyengar, L., and Rao, S.P. (1977). Mechanism of disinfection: Effect of chlorine on cell membrane functions. *Water Res.* 11:727–729.

Wagner, W.F., Gump, R.J., and Hart, E.N. (1952). Factors affecting the stability of aqueous potassium ferrate (VI) solutions. *Anal. Chem.* 24:1497–1498.

Waite, T.D. (1978a). *Management of waste water residuals with iron (VI) ferrate.* First Annual Report Grant ENV 76-83897. Washington, DC: National Science Foundation.

Waite, T.D. (1978b). Inactivation of *Salmonella* sp, *Shigella* sp, *Streptococcus* sp and f2 virus by iron (VI) ferrate. Paper 33-4 presented at the Annual Meeting of the American Water Works Association, Atlantic City, NJ.

Waite, T.D., and Gray, K.A. (1984). Oxidation and coagulation of wastewater effluent utilizing ferrate VI ion. *Stud. Environ. Sci.* 23:407–420.

Waller, K., Swan, S.H., DeLorenze, G., and Hopkins, B. (1998). Trihalomethanes in drinking water and spontaneous abortion. *Epidemiology* 9:134–140.

Wardle, M.D., and Reninger, G.M. (1975). Bactericidal effect of hydrogen peroxide on spacecrafts isolates. *Appl. Microbiol.* 30(4):710–711.

Wei, I.W. (1972). Chlorine ammonia breakpoint reactions: Kinetics and mechanism. Ph.D. dissertation, Harvard University, Cambridge, MA.

Wei, W., and Morris, J.C. (1974). Dynamics of breakpoint chlorination. In A.J. Rubin (ed.), *Chemistry of water supply, treatment, and distribution.* Ann Arbor, MI: Ann Arbor Science Publishers, pp. 297–332.

Weidenkopf, S.I. (1958). Inactivation of type I poliomyelitis virus with chlorine. *Virology* 5:5–7.

Welch, W.A. (1963). Potassium permanganate in water treatment. *J. Am. Water Works Assoc.* 55:735–741.

White, G.C. (1972). *Handbook of chlorination for potable water, wastewater, cooling water, industrial processes and swimming pools.* New York: Van Nostrand Reinhold Company.

Wood, R.H. (1958). The heat, free energy and entropy of the ferrate (VI) ion. *J. Am. Chem. Soc.* 80:2038–2041.

World Health Organisation. (1993). *Guidelines for drinking-water quality: Recommendations,* 2nd ed., vol. 1. WHO.

Wright, J.M., Schwartz, J., Vartiainen, T., Maki-Paakkanen, J., Altshul, L., and Harrington, J.J. (2002). 3-Chloro-4-(dichloromethyl)-5-hydroxy-2(5H)-furanone (MX) and mutagenic activity in Massachusetts drinking water. *Environ. Health Perspect.* 110:157–164.

Wyss, O., and Stockton, J.R. (1947). The germicidal action of bromine. *Arch. Biochem.* 12:267–271.

Xu, J., Dozier, A., and Bhattacharyya, D. (2005). Synthesis of nanoscale bimetallic particles in polyelectrolyte membrane matrix for reductive transformation of halogenated organic compounds. *J. Nanopart. Res.* 7:449.

Yasinskii, A.V., and Kuznetsova, V.F. (1973). Disinfection of water containing vibrios by silver ions. *Aktual. Vopr. Sanit. Microbiol.* 112–113.

Yates, M.V., Malley, J., Rochelle, P., and Hoffman, R. (2006). Effect of adenovirus resistance on UV disinfection requirements: A report on the state of adenovirus science. *J. Am. Water.* 98(6):93–106.

Yoshpe-Purer, Y., and Eylan, E. (1968). Disinfection of water by hydrogen peroxide. *Health Lab Sci.* 5:233–238.

You, Y.W., Han, J., Chiu, P.C., and Jin, Y. (2005). Removal and inactivation of waterborne viruses using zerovalent iron. *Environ. Sci. Technol.* 39(23):9263–9269.

Zhdanov, Y.A., and Pustovarova, O.A. (1967). Oxidation of alcohol and aldehydes by potassium ferrate. *Zn. Onsch. Khim.* 37(12):2780.

Zimmermann, W. (1952). Oliodynamic silver actions. I. The action mechanism. *Z. Hyg. Infektionskr.* 135:403–413.

3

Physical Disinfection

3.1 Introduction

Conventional drinking water disinfectants such as chlorine and other chemical oxidants have successfully protected public health against waterborne microbial diseases for decades since ensuring the microbiological safety of drinking water is of paramount importance worldwide. This has been elaborately described in Chapter 2. However, chemical disinfection has several demerits, including the production of potentially toxic by-products, problems of taste and odour, and resistance of certain emerging pathogens such as *Giardia* and *Cryptosporidium* species, to name a few. This has led to a reappraisal of traditional chemical disinfection practice. Effects of climate change leading to natural disasters like flood and drought could exacerbate these concerns, which could make treatment more challenging. In the longer term, the pressure on the chemical industry to reduce production of chlorine and chlorine-based disinfectants for environmental reasons (greener routes) may force water companies to turn to other disinfectants or employ non-chemical disinfection processes.

A number of commercially available alternative processes, such as membrane processes, are able to remove bacteria, viruses, and protozoa, as well as a range of chemical contaminants. These are coming into use, but generally only on a small scale. It may be possible to operate these processes with no chemical disinfection, or at least to reduce the amount of chemicals used for the final disinfection. A range of pressure-driven membrane processes—microfiltration, ultrafiltration, nanofiltration, and reverse osmosis, in order of decreasing pore size or increasing operating pressures—are also capable of disinfection, as well as the removal of chemical contaminants, depending on pore size. The use of membrane processes would avoid the formation of disinfection by-products and would reduce the concentrations of other undesirable chemicals, giving a net benefit in terms of toxicological issues. The main microbiological concerns with membrane systems are ensuring the integrity of the membrane and monitoring the efficiency of microorganism removal; with conventional chlorination the residual chlorine concentration is easily monitored and provides reassurance that disinfection has been carried out effectively. Moreover, over a period of time the microbes from the source water start getting accumulated on the surface of the membrane, leading to its biofouling. Often, membranes have to be replaced, thereby affecting the economics of the water treatment. In addition, the membrane cost is high, whereas the established chemical disinfection methods are cheaper. If the water to be disinfected has particulate matter and other impurities, it may lead to irreversible clogging of the membrane. Hence, pretreatment like filtration is normally required before the membrane can be used. Alternatives to chemical disinfection, such as UV irradiation, are also being used for disinfection of drinking water. UV is capable of inactivating bacteria and viruses, and possibly protozoan parasites. Such nonconventional processes and disinfection methods could in principle be used to replace, or at least greatly reduce, the use of chemical disinfection of drinking water. This chapter describes these and other physical processes that can replace or reduce chemical disinfection in the operation of water disinfection.

3.2 Ultraviolet Radiation

UV radiation energy waves are the range of electromagnetic waves 100 to 400 nm long (between the x-ray and visible light spectrums). The division of UV radiation may be classified as vacuum UV (100–200 nm), UVC (200–280 nm), UVB (280–315 nm), and UVA (315–400 nm). In terms of germicidal effects, the optimum UV range is between 245 and 285 nm. UV radiation quickly dissipates into water to be absorbed or reflected off material within the water. As a result, no residual is produced. This process is attractive from a disinfection by-product (DBP) formation standpoint; however, a secondary chemical disinfectant is required to maintain a residual throughout the distribution system, which may be subjected to recontamination. UV disinfection utilises either low-pressure lamps that emit maximum energy output at a wavelength of 253.7 nm, medium-pressure lamps that emit energy at wavelengths from 180 to 1370 nm, or lamps that emit at other wavelengths in a high-intensity "pulsed" manner.

3.2.1 Generation of UV

The lamps typically used in UV disinfection consist of a quartz tube filled with an inert gas, such as argon, and small quantities of mercury. UV lamps operate in a similar fashion as fluorescent lamps. UV radiation is emitted from electron flow through ionised mercury vapour to produce UV rays with sufficient energy required for disinfection. The difference between the two lamps is that the fluorescent lamp bulb is coated with phosphorous, which converts the UV radiation to visible light, whereas the UV lamp is not coated, so it transmits the UV radiation generated by the arc (White, 1992). The intensity of medium-pressure lamps is much greater than that of low-pressure lamps. Thus, fewer medium-pressure lamps are required for an equivalent dosage. For small systems, the medium-pressure system may consist of a single lamp. Although both types of lamps work equally well for inactivation of organisms, low-pressure UV lamps are recommended for small systems because of the reliability associated with multiple low-pressure lamps (DeMers and Renner, 1992) as opposed to a single medium-pressure lamp, and for adequate operation during cleaning cycles. Typically, low-pressure lamps are enclosed in a quartz sleeve to separate the water from the lamp surface. This arrangement is required to maintain the lamp surface operating temperature near its optimum of 40°C. Teflon sleeves are an alternative to quartz sleeves; however, they absorb 35% of UV radiation compared to quartz sleeves, which absorb only 5% (Combs and McGuire, 1989). Therefore, Teflon sleeves are not recommended.

For drinking water applications two types of UV reactors, closed vessel and the open channel, are conventionally used. The former is recommended by the U.S. Environmental Protection Agency (USEPA) due to its merits, such as smaller footprint, small floor space requirement, less pollution from airborne material, limited external contamination, minimal personnel exposure to UV rays, and simplicity of design, to name a few (USEPA, 1996). A conventional closed-vessel UV reactor is capable of providing UV dosages adequate to inactivate bacteria and viruses. However, it is incapable of the higher dosages required for protozoan cysts. To increase the dosage, either the number of UV lamps or the exposure time should be increased. Additional design features, such as UV sensors, alarms, automatic cleaning cycles, and so forth, are also standard components of some systems.

Microscreening/UV and pulsed UV are systems that claim to inactivate *Giardia* cysts and *Cryptosporidium* oocysts, which are generally regarded to be resistant to conventional UV treatment. Microscreening chambers were designed in the late 1990s and essentially consist of two treatment chambers, each having a 2 μm porosity metal screen. Low-pressure mercury lamps, three in number, are present on each side of the screen. As water containing oocysts passes through the unit, oocysts are captured in the first screen, where they are exposed to a preset UV dose. The oocysts on the first screen are back flushed by reversing the flow within the unit to reach the second screen, where the oocysts are again trapped and exposed to yet another preset UV dose (Clancy et al., 1997). Johnson (1997) stated that such a system is capable of achieving total UV doses of 8000 mW.s/cm^2, sufficient to inactivate *Giardia* cysts and *Cryptosporidium* oocysts. Figure 3.1 shows the working of a microscreening/UV chamber.

Pulsed UV is yet another remarkable methodology to inactivate microorganisms. The pulsed UV reactor consists of a flash chamber typically 2 in. in diameter that is fitted with xenon flash tubes in the center of the chamber. Capacitors used in the unit are designed to build up electricity and deliver it

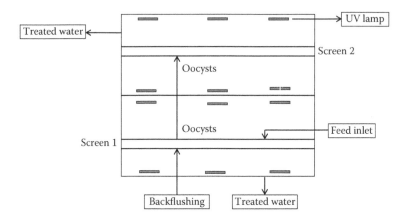

FIGURE 3.1
Microscreening/UV.

in pulses, hence the name pulsed UV. As water to be treated flows through this chamber, with each pulse, the flash tubes give off high-intensity, broadband radiation, including germicidal UV radiation, which irradiates the flowing water with irradiances of 75 mW.s/cm^2 at 2 cm from the flash tube surface (Clancy et al., 1997). The UV dose can be adjusted by increasing or decreasing the frequency and duration of the pulsing.

Recently, GaN-based ultraviolet C (UVC) light-emitting diodes (LEDs) have been investigated for water disinfection. They offer significant advantages compared to conventional mercury lamps due to their compact form factor, low power requirements, high efficiency, nontoxicity, and overall robustness. However, these devices still suffer from low emission power and relatively short lifetimes. In a very recent report, GaN-based UV LEDs could very effectively inactivate *Bacillus subtilis* spores in water samples with varying qualities. It appears to be a promising alternative for water disinfection, especially for decentralised and mobile systems (Würtele et al., 2011).

3.2.2 Mechanism of Bacterial Inactivation

Electromagnetic radiation in the wavelengths ranging from 240 to 280 nm effectively inactivates microorganisms by irreparably damaging their nucleic acid. The most potent wavelength for damaging deoxyribonucleic acid (DNA) is approximately 254 nm (Wolfe, 1990), and microorganism nucleic acids absorb light energy in the wavelength of 240 to 280 nm (Jagger, 1967). The germicidal effects of UV light involve photochemical damage to RNA and DNA within the microorganisms. DNA and RNA carry genetic information necessary for reproduction; therefore, damage to either of these substances can effectively sterilise the organism. The organism also stops its other functions, like protein synthesis, required for its survival. The mechanism of the damage can be explained as follows: Three primary types of pyrimidine molecules are cytosine (found in both DNA and RNA), thymine (found only in DNA), and uracil (found only in RNA). UV irradiation causes dimerisation of these pyrimidine molecules, which leads to distortion of the DNA helical structure. Once this happens, replication of the nucleic acid becomes very difficult (Snider et al., 1991). It is indeed interesting to note that the process does not end here.

Reactivation, an amazing phenomenon of a natural self-defense mechanism, has evolved in microorganisms for millions of years. With the aid of this spectacular natural process, the microbes are able to reverse the DNA damage caused by UV radiation. Fascinatingly, the microbes bring about reactivation by two modes: photoreactivation and dark repair.

Photoreactivation occurs in the presence of visible light by an enzyme called photolyase, which reverses the UV-induced damage. On the other hand, dark repair happens in the absence of light with the aid of a myriad of complex combinations of enzymes. One common feature in both forms of repair is the necessity of activation of the enzymes. While the visible light activates photolyase, nutrients present with the cell are generally sufficient to accomplish the same task. This results in the damaged DNA being repaired, thus allowing replication to occur. The extent of reactivation varies among organisms. Coliform indicator organisms and some bacterial pathogens such as *Shigella* have exhibited the photoreactivation mechanism; however, viruses and other types of bacteria cannot photoreactivate (USEPA, 1980, 1986; Hazen and Sawyer, 1992). Moreover, DNA damage tends to become irreversible over time. Therefore, there is a critical period during which photoreactivation can occur. To minimise the effect of photoreactivation, UV contactors should be designed to either shield the process stream or limit the exposure of the disinfected water to sunlight immediately following disinfection (Tchobanoglous, 1997). In general, UV rays can be used to achieve either partial or complete disinfection. The former is preferred most of the time, and the degree of disinfection is governed by the permissible limits of the microorganism being treated laid down by the water authorities. Recent investigation of the dark repair mechanism in environmental isolated *Pseudomonas* and *Enterococcus* revealed that the key induction mechanism was the recA system, the presence of which was analysed and confirmed by the researchers. It appeared that the recA protein inactivation depended on the incubation time after UV treatment, which differed among various drinking waters, as well as opportunistic bacteria (Jungfer et al., 2007). Subsequent research led to the complete genome sequencing and analysis of a large number of eubacteria and archeobacteria, which revealed the genes encoding enzymes involved in the repair pathways. Comparison of the homologous sequences between the different species showed how the repair system has been transmitted during evolution (Goosen and Moolenaar, 2008).

3.2.3 Disinfection Efficacy

UV disinfection has been determined to be adequate for inactivating bacteria and viruses. Most bacteria and viruses require relatively low UV dosages for inactivation, typically in the range of 2 to 6 mW.s/cm^2 for a 10-fold reduction. Protozoan oocysts, in particular *Giardia* and *Cryptosporidium*, are considerably more resistant to UV inactivation than other microorganisms. Results of several studies investigating the ability of UV to inactivate bacteria, viruses, and protozoa are described in the following sections.

3.2.3.1 Bacteria and Virus Inactivation

UV doses required for bacteria and virus inactivation are relatively low. One study determined that UV was comparable to chlorination for inactivation of heterotrophic plate count bacteria following treatment using granular activated carbon (der Kooij, Hijnen, and Kruithof, 1989). A study of the ability of UV and free chlorine to disinfect a virus-containing groundwater showed that UV is a more potent virucide than free chlorine, even after the chlorine residual was increased to 1.25 mg/l at a contact time of 18 min (Slade et al., 1986). The UV dose used in this study was 25 mW.s/cm^2. Several studies have been undertaken to understand the extent of UV influence required for various classes of microorganisms found in water. It has been noted that the environmental strains and spores are generally more resistant than the laboratory grown strains. Therefore, the minimum inhibitory concentration (MIC) of UV fluence requires a correction factor of 2 or 3 in its calculation (Slade et al., 1986). This aspect has not been investigated very critically for viruses and protozoa. In another study, the presence of naturally occurring particles in surface water was found to protect the indigenous coliforms found in four surface waters from UV irradiation and hamper the disinfection efficacy. The results suggest that particles as small as 11 μm, naturally found in surface water with low turbidity (<3 nephelometric turbidity units [NTUs]), are able to harbor indigenous coliform bacteria and offer protection from low-pressure UV light (Cantwell and Hofmann, 2008).

3.2.3.2 Protozoa Inactivation

Even though protozoa were once considered resistant to UV radiation, recent studies have shown that ultraviolet light is capable of inactivating protozoan parasites. However, results indicate that these organisms require a much higher dose than that needed to inactivate other pathogens. Less than 80% of *Giardia lamblia* cysts were inactivated at UV dosages of 63 mW·s/cm^2 (Rice and Hoff, 1981). A 10-fold reduction of *Giardia muris* cysts was obtained when the UV dose was increased to 82 mW·s/cm^2 (Carlson, 1985). To achieve 20-fold inactivation of *Giardia muris* cysts, a minimum ultraviolet light dose of above 121 mW·s/cm^2 is needed. Karanis et al. (1992) examined the disinfection capabilities of ultraviolet light against *Giardia lamblia* cysts extracted from both animals and humans. Both groups suffered a 100-fold reduction at UV doses of 180 mW·s/cm². Two important factors to consider when determining dose requirements for *Giardia* inactivation are the parasite source and the growth stage of the microorganism (Karanis et al., 1992). Results from recent studies show a potential for inactivating *Cryptosporidium parvum* oocysts using ultraviolet light disinfection. A 100- to 1000-fold reduction in the viability of *Cryptosporidium parvum* oocysts was achieved using a low-pressure ultraviolet light system with a theoretical minimum intensity of 14.58 mW/cm² and a contact time of 10 min (ultraviolet dose of 8748 mW·s/cm²) (Campbell et al., 1995). The combination filter-UV system described by Johnson (1997) is capable of delivering doses as high as 8000 mW·s/cm², sufficient to achieve 2-log *Cryptosporidium* oocyst inactivation. A pulsed UV process that delivered a minimum dose of 1900 mWs/cm² to any particle within the reactor was found to achieve *Cryptosporidium* oocyst inactivation levels in the range of 100-fold.

In the study by Clancy et al. (1997), the reactor residence time was 4.7 s and the unit was operated to deliver 46.5 pulses per volume (with a pulse frequency of 10 Hz). Each pulse transfers power at the intensity rate of about 41 mWs/cm².

Research indicates that when microorganisms are exposed to UV radiation, a constant fraction of the living population is inactivated during each progressive increment in time. This dose-response relationship for germicidal effect indicates that high-intensity UV energy over a short period of time would provide the same kill as a lower-intensity UV energy at a proportionally longer period of time. This is a typical first-order response where concentration and time can be exchanged and the final product of the two will decide the final extent of disinfection. The UV dose required for effective inactivation is determined by site-specific data relating to the water quality and log removal required (extent of disinfection required). Based on first-order kinetics, the survival of microorganisms can be calculated as a function of dose and contact time (USEPA, 1996; White, 1992). For high rate of removal, the remaining concentration (intact microorganisms) of organisms appears to be solely related to the dose and water quality, and not dependent on the initial microorganism density. Tchobanoglous (1997) suggested the following relationship between coliform survival and UV dose:

$$N = f \cdot Dn$$

where:
N = Effluent coliform density, per 100 ml
D = UV dose, mW.s/cm²
n = Empirical coefficient related to dose and the geometry of the system
f = Empirical water quality factor (usually > 1)

The empirical water quality factor reflects the presence of particles, color, and so forth, in the water. For water treatment, the water quality factor is expected to be a function of turbidity and transmittance (or absorbance).

Based on the available research literature, it appears that although exceptional for disinfection of small microorganisms such as bacteria and viruses, UV doses required to inactivate larger protozoa such as *Giardia* and *Cryptosporidium* are several times higher than for bacteria and virus inactivation (DeMers and Renner, 1992; White, 1992). As a result, UV is often considered in concert with ozone or hydrogen peroxide to enhance the disinfection effectiveness of UV or for groundwater, where *Giardia* and *Cryptosporidium* are not expected to occur. Since UV radiation is energy in the form of electromagnetic waves, its effectiveness is not limited by chemical water quality parameters. For instance, it appears that pH, temperature, alkalinity, and total inorganic carbon do not impact the overall effectiveness of UV disinfection (AWWA and ASCE, 1990). However, calcium hardness may cause problems for keeping the lamp sleeves clean and functional. The presence, or addition, of oxidants (e.g., ozone or hydrogen peroxide) enhances UV radiation effectiveness. The presence of some dissolved or suspended matter may shield microorganisms from the UV radiation. For instance, iron, sulphites, nitrites, and phenols all absorb UV light (DeMers and Renner, 1992). Salient factors that affect the disinfection efficiency of UV are discussed in Table 3.1.

3.2.4 Disinfection By-Products of UV Radiation

Continuous-wave UV radiation at doses and wavelengths typically employed in drinking water applications does not significantly change the chemistry of water, nor does it significantly interact with any of the chemicals within the water (USEPA, 1996). Therefore, no natural physiochemical features of the water are changed and no chemical agents are introduced into the water. In addition, UV radiation does not produce a residual. As a result, formation of trihalomethane (THM) or other DBPs with UV disinfection is minimal. Unlike other disinfectants, UV does not inactivate microorganisms by chemical reaction. However, UV radiation causes a photochemical reaction in the organism RNA and DNA. Literature suggests that UV radiation of water can result in the formation of ozone or radical oxidants (Ellis and Wells, 1941; Murov, 1973). Because of this reaction, a lot of research was directed in determining whether UV forms similar by-products to those formed by ozonation or other advanced oxidation processes. Groundwater samples analysed for aldehydes and ketones before and after UV irradiation revealed positive results only for 1 out of 20 samples analysed. Low levels of formaldehyde were also measured in this case. No significant DBP formation was observed on subsequent chlorination or by varying dosage of UV. Interestingly, surface water studies revealed that low levels of formaldehyde were produced upon UV irradiation. This was attributed to the presence of humic substances in water (Malley et al., 1995). The DBP formation rate studies indicated that UV

TABLE 3.1

Factors Affecting UV Disinfection

Serial No.	Factor	Effect on UV Disinfection Efficiency	Reference
1	Chemical and biological films that develop on the surface of UV lamps	Decrease UV intensity reaching water and thereby reduce the efficiency.	DeMers and Renner (1992)
2	Reactor geometry	Reactor geometry that creates dead space between the UV contactors and the microbes can lead to decrease in efficiency.	Hazen and Sawyer (1992)
3	Short circuiting	UV systems typically provide contact times of the order of seconds. Therefore, it is extremely important that the system configuration limit the extent of short circuiting and presence of zones of UV shadows.	DeMers and Renner (1992)
4	Microorganism clumping	They can shade some microbes from being exposed to UV, thereby reducing the disinfection efficiency.	Hazen and Sawyer (1992)
5	Turbidity	Increasing turbidity can lead to lower efficiency due to reduced UV rays reaching the microorganisms.	Yip and Konasewich (1972)

radiation did not significantly affect DBP formation rates when chlorine or chloramines were used as the post-UV chemical disinfectant. Continuing research in this area has revealed contrasting but interesting results. Very recently, the effect of UV treatment on the dissolved organic matter (DOM) and its structure revealed that UV irradiation can cleave DOM, but the molecular weight of the broken DOM is not low enough to be used directly as a food/substrate by microbes in the distribution system. It has been reported that the changed DOM structure can affect the water quality of the distribution system, as it can increase the chlorine demands and DBP formation by postchlorination (Choi and Choi, 2010).

Thus, the primary use of UV radiation appears to be to inactivate pathogens to regulated levels. It is effective against most viruses, spores, and cysts and does not require chemical generation, handling, transport, or storage of hazardous chemicals. Lack of residual effects is favourable for human consumption and safety for aquatic life. It can be regarded as a user-friendly process with shorter contact times and less space for equipment. On the other hand, reactivation by microbes due to its self-defense, fouling of UV tubes, presence of turbidity, and the cost are a few of the demerits of UV disinfection.

3.3 Solar Disinfection

Solar disinfection or use of sunlight for water disinfection is an ancient tradition that has been practiced for centuries. Stone Age man effectively harnessed solar energy for rendering water clean of microbes. However, the scientific reason behind the phenomenon was not well comprehended. Over a period of time modern-age man deciphered the scientific basis for disinfection by solar rays. The journey is an interesting one, beginning as early as the late 1880s when the first systematic studies were initiated by Downes and Blunt. Sunlight, especially short-wavelength radiation, effectively prevented bacterial growth in nutrient broth, a routine microbiological media used to culture microbes (Downes and Blunt, 1987). Although research continued to progress in this arena, it was only in 1980 that solar radiation was reported for water disinfection. This was regarded as a low-cost, simple technique for providing safe drinking water in developing countries with sunny climates. Solar disinfection involves filling a transparent glass or plastic bottle with the water to be treated and exposing it to sunlight for several hours. Myriad microbes, including indicators of pollution such as *Escherichia coli*, faecal coliforms, as well as pathogenic microorganisms like *Salmonella typhi* and *Shigella flexneri*, several yeasts, and moulds are successfully eliminated by solar disinfection.

3.3.1 Mechanisms of Solar Disinfection

Sunlight brings about disinfection by direct as well as indirect means. The solar rays can be directly absorbed by the microbial cell, damaging the DNA and thereby causing cell death. Indirectly, the solar rays may excite photosensitiser molecules such as porphyrins and pigments (Curtis et al., 1992) that are commonly present within the cell, which in turn may trigger a type I or type II mechanism. The former is essentially a reaction of the excited photosensitiser with cellular biomolecules, and the latter, more common mechanism is the reaction of the excited photosensitiser with molecular oxygen. Eventually this leads to the formation of reactive oxygen species (ROS) such as hydroxyl radicals, which damage the cellular membrane, proteins, and DNA (Jeffrey et al., 1996).

Yet another mechanism by which solar disinfection is achieved is due to the thermal effect of solar radiation. Here, the absorption of sunlight, especially solar infrared radiation, raises the temperature of the water to a point where microbes are inactivated, in a process often termed *solar pasteurisation*, by analogy with commercial pasteurisation. A simple batch-process solar pasteurisation has been carried out by using a solar box cooker and black-painted container, showing that faecal coliforms are inactivated at water temperatures of 60°C or greater (Ciochetti and Metcalfe, 1984). The overall mechanisms involved in solar disinfection have been illustrated in Figure 3.2.

Several research studies have reported a synergy between optical and thermal inactivation. Thus, Tyrell (1976) showed a synergistic effect of UV radiation and heat in the inactivation of *E. coli*, and Wegelin et al. (1994) subsequently demonstrated that temperatures above 50°C result in a threefold decrease in the UVA radiation dose required for inactivating *E. coli*, with even more striking effects for bacteriophages and enteroviruses. McGuigan et al. (1998) showed synergy at temperatures above 45°C, where the combined effects of simulated sunlight and heat resulted in a greater rate of inactivation of *E. coli* than that predicted from the rates obtained by using each

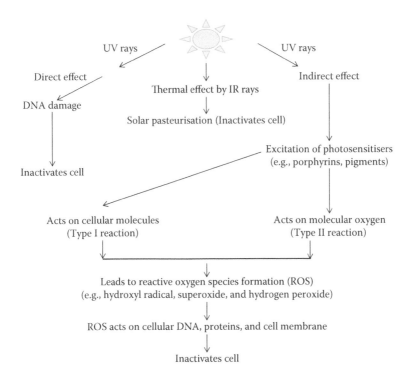

FIGURE 3.2
Mechanisms of solar disinfection.

factor in isolation. Kehoe et al. (2001) have shown that the inactivation of *E. coli* can be enhanced by a factor of almost twofold by adding an aluminium foil backing to the containers to reflect UV and visible light, thereby enhancing the effect of the optical component of the process.

3.3.2 Factors Affecting Efficacy of Solar Disinfection

The efficacy of solar disinfection is dependent on several factors. Broadly, they may be classified as chemical aspects that include the presence of organic and inorganic compounds and dissolved oxygen in the water to be treated, physical aspects like the quality of light and its intensity, temperature, type of container, and so forth, and the microbiological aspect. The factors affecting solar disinfection efficacy are discussed in Table 3.2.

TABLE 3.2

Factors That Affect the Efficacy of Solar Disinfection

Serial No.	Factor	Effect on Solar Disinfection Efficacy	Reference
1	Organic compounds (e.g., humic acid)	Dual effect. May act as a photosensitiser and enhance solar disinfection by ROS production. On the other hand, it may absorb solar rays and thereby reduce optical inactivation.	Davies and Evison (1991)
2	Inorganic compounds (e.g., high concentration of salts)	Acts synergistically with solar rays and promotes inactivation of microbes. Indirectly, increases the pH to alkaline range, which leads to cell membrane damage.	Vicars (1999)
3	Dissolved oxygen	Solar disinfection efficacy is enhanced considerably in the presence of dissolved oxygen. This may be attributed to photooxidation with production of ROS.	Reed et al. (2000)
4	Light quality and intensity	290–350 nm, corresponding to the UVB and UVA regions of the spectrum, is reported to cause solar water disinfection. Effective solar disinfection requires around 3–5 h of strong sunlight at an intensity above 500 W/m^2.	Acra et al. (1984), Oates et al. (2003)
5	Temperature	At temperatures above 40–50°C a synergistic effect of thermal and optical inactivation is observed. Lower temperatures have no significant effect.	Wegelin et al. (1994)

TABLE 3.2

Factors That Affect the Efficacy of Solar Disinfection

Serial No.	Factor	Effect on Solar Disinfection Efficacy	Reference
6	Type of container	Clear glass and polyethylene terephathalate (PET) bottles are generally recommended for solar disinfection.	Acra et al. (1984), Conroy et al. (1996, 1999)
7	Turbidity	Bacterial inactivation decreases as the turbidity is raised to around 300 NTU. Interestingly, turbid waters (>200 NTU) absorb visible and IR rays due to the presence of suspended particles and are reported to lead to complete disinfection of water only due to the thermal effect of solar disinfection.	Acra et al. (1984, 1990), Wegelin et al. (1994), Kehoe et al. (2001), Joyce et al. (1996)
8	Type of microbe to be inactivated and its growth phase	Faecal indicators like *E. coli* exhibit an initial delay of 0.5–2 (pronounced in stationary phase cells) h, after which the solar inactivation is exponential. Viruses require twice the dose of solar rays compared to bacteria. Fungi require between 3 and 6 h for inactivation depending on the species irradiated. Cysts are rendered inactive by thermal effects, i.e., within 10 min at 56°C.	Reed (1997b), Wegelin et al. (1994), Acra et al. (1984), Ciochetti and Metcalfe (1984)

3.3.3 Inactivation of Microorganisms by Solar Radiation

The most extensive studies of the dynamics of solar disinfection have been carried out by using bacteria, especially with pure cultures of the faecal indicator bacterium *E. coli*. Typical inactivation curves show an exponential decrease in the bacterial count against time, often with an initial shoulder or plateau, lasting 0.5 to 2 h, corresponding to a delay in the inactivation process. This shoulder is most marked in stationary phase cells (Reed, 1997b) and is generally interpreted in terms of a multiple-target model of inactivation (Davies-Colley, Donnison, and Speed, 1997; Wegelin et al., 1994). After this initial shoulder, the inactivation kinetics generally follows a single-exponential decay function, giving a straight line on a log-linear graph (e.g., Reed, 1997b; Wegelin et al., 1994). However, this is not always the case, and McGuigan et al. (1998) have described solar inactivation kinetics of a Kenyan isolate of *E. coli* in terms of a double-exponential decay function involving a light-sensitive and a light-resistant subpopulation that gives a nonlinear relationship, especially at high irradiances. Salih (2002) has also proposed a

TABLE 3.3

T_{90} Values of Bacteria Exposed to Solar Radiation

Bacteria	T_{90} (min)	Reference
Escherichia coli	38	Acra et al. (1990)
Faecal coliforms	70	Reed (1997a,b)
Faecal streptococci	65	Reed (1997a,b)
Shigella flexneri	67	Kehoe et al. (2001)
Vibrio cholerae	171	Kehoe et al. (2001)

more complex model based on the combined effects of (1) exposure and (2) bacterial load. A widely used means of representing the exponential decay component of the inactivation process is to calculate the T_{90} value (i.e., the time required to reduce the plate count by 90%) (Guillard et al., 1997). Although no single value applies to a particular bacterium, most of the T_{90} values lie between 30 and 120 min. Of course, such T_{90} values will be influenced by the various chemical and physical factors listed above, and there may also be between-batch variability when experiments are repeated (Davies-Colley et al., 1997). However, Table 3.3 provides an indication that a full day of sunlight should be sufficient to inactivate 99.9% of all of the bacteria listed, even with an initial lag period of 1 to 2 h.

There is considerably less information on the inactivation of other microbes. Wegelin et al. (1994) have shown that coliphage f2 and bovine rotavirus are inactivated by a similar amount of light to that required for *E. coli*, whereas encephalomyocarditis virus needed twice the dose, demonstrating the virucidal effects of natural and simulated sunlight. Acra et al. (1984) report inactivation of a range of fungi, including *Aspergillus niger, Aspergillus flavus, Candida* sp., and *Geotrichum* sp. within 3 h, and *Penicillium* sp. within 6 to 8 h. Cysts of *Giardia* spp. and *Entamoebahistolytica* can be inactivated by the thermal effects of sunlight (e.g., within 10 min at 56°C) (Ciochetti and Metcalfe, 1984), though such temperatures are only readily achieved in prolonged strong sunlight under conditions in which the thermal effects are boosted (e.g., by using absorptive black surfaces or a solar hot box cooker). Optical effects alone are unlikely to inactivate protozoan cysts, as shown for *Acanthamoebapolyphaga* (Lonnen et al., 2004).

Thus, solar disinfection is not universally applicable, but it may be appropriate under circumstances in which there is no realistic alternative treatment process. There are a myriad of instances where this could be possible. Several rural areas and villages have access only to sewage-contaminated water for potable use. Solar disinfection can be implemented as a medium- to long-term water disinfection in such areas. Inability for widespread distribution of piped water supply to a rural population can also be an avenue for solar disinfection, especially when chemical disinfectants become unaffordable for the lower economic strata population. Apart from this, solar disinfection can be looked upon as an effective alternative for short-term water disinfection in emergencies such as war, natural calamities like floods, and epidemic conditions where there is contamination due to a specific pathogen like *Vibrio chlorea*, in the case of cholera (Anon., 2002; Ribeiro, 2000).The successful use of solar disinfection in reducing the incidence of cholera during field trials in Kenya (Conroy et al., 2001) confirms the practical value of this approach. The World Health Organisation has advocated solar disinfection due to its immense benefits (Anon., 2001; Sobsey, 2002). The pathogenic protozoan parasite *Cryptosporidium parvum* associated with cryptosporidiosis has been a major challenge for all water disinfection techniques and solar disinfection (SODIS) is no exception. Research in this area includes mathematical models to investigate the combined effects of solar radiation intensity in the range of 320 nm to 10 µm, water turbidity in the range of 5 to 300 NTU, and exposure time between 4 and 12 h on the viability of the *C. parvum* oocycts. It has been reported that all the above three factors had a major significance on the oocycts' survival, with the greatest effect being the intensity of radiation. Levels of ≥ 600 W/m^2 and times of exposure between 8 and 12 h were required to reduce the oocyst infectivity in water samples with different degrees of turbidity (Gómez-Couso et al., 2009). Yet another interesting investigation of the antimicrobial activity of SODIS was carried out recently where simulated SODIS in the presence and the absence of riboflavin (250 µM) was investigated for the inactivation of various protozoa and helminthes, such as *Acanthamoeba*, *Naegleria*, *Entamoeba*, and *Giardia*. SODIS at an optical irradiance of 550 W/m^2 for up to 6 h resulted in significant inactivation of these organisms. Although the addition of riboflavin significantly increased the level of inactivation observed with cysts of *A. castellanii*, *Cryptosporidium* oocysts, and *Ascaris* ova, exposure to SODIS in the presence and absence of riboflavin for 6 to 8 h resulted in a negligible reduction in the viability of both organisms (Heaselgrave and Kilvington, 2011).

Since ROS formation by excited photosensitisers forms the basis of optical disinfection by solar irradiation, a lot of research was directed toward the addition of extraneous agents, which could play the role of photosensitisers. In the late 1990s, researchers added simple dyes such as methylene blue or rosebengal to enhance the production of ROS in aqueous solution and thereby increase the antimicrobial effects of light (e.g., Chilvers, Reed, and Perry, 1999; Wegelin, 1994). The major concern was the consumption of water containing these dyes. This was eventually addressed (Acra and Ayoub, 1997) by adding chlorine to decolourise the dye prior to the consumption of the treated water, although solar radiation can also fade colours and therefore reduce the colour given by the dye. However, this process was found more appropriate for wastewater treatment due to the impractical in-field condition for potable water production in developing countries where the availability of two chemicals, the dye and chlorine, was relatively difficult.

Continuing efforts in the search for effective photosensitisers led the researchers to titanium dioxide (TiO$_2$). It could be used as a stable photosensitiser since excitation of TiO_2 by short-wavelength light (<385 nm) leads to the generation of ROS, principally hydroxyl radicals (Harper et al., 2001; Ollis et al., 1991). The literature reports initial studies on water containing the gram-negative bacterium *E. coli*, the gram-positive bacterium *Lactobacillus acidophilus*, or the yeast *Saccharomyces cerevisiae*. These experiments revealed that water loaded with platinum and TiO_2 could be effectively disinfected by using a metal halide lamp within 1 to 2 h (Matsunaga et al., 1985). Photocatalytic inactivation has also been reported for several other microbes, including bacteria, fungi, viruses, and some protozoa, as shown in Table 3.4. Thus, conventional batch-process solar disinfection can be modified to take advantage of TiO_2-enhanced photocatalysis, thereby reducing the irradiation dose (Duffy et al., 2003) and extending the range of microbes against which it is effective (Lonnen et al., 2005). In one study, where SODIS and solar photocatalytic disinfection were compared for inactivation of waterborne protozoan, fungal, and bacterial microbes, it was observed that after 8 h simulated solar exposure at an intensity of 870 W/m^2 in the 300 nm to 10 μm range, and 200 W/m^2 in the 300 to 400 nm UV range, both techniques achieved 4 log reduction in the microbes investigated. However, both methods were reported to be ineffective against the cysts of *A. polyphaga* (Lonnen et al., 2005). Eventually, more efforts were directed toward the disinfection of *Cryptosporidium* oocycts, a major waterborne pathogen. In yet another study, SODIS reactors fitted with flexible plastic inserts coated with TiO_2 powder were found to be much more effective against *Cryptosporidium parvum* oocyst than those that were not coated, indicating its promising application in household water disinfection (Méndez-Hermida et al., 2007).

Factors that affect TiO_2-assisted photocatalytic solar disinfection are many. Alkaline conditions and the presence of organic or inorganic substances, including disinfection by-products, decrease the effectiveness of disinfection. These substances have a tendency to deposit on the surface of the catalyst, thereby reducing its ability to absorb UV rays and inactivate microorganisms. Studies demonstrate that the presence of catalase enzyme inhibited the effectiveness of TiO_2-assisted photocatalytic solar disinfection. This revealed the direct role of hydroxyl radicals in the inactivation process since catalyse

TABLE 3.4

Microorganisms Inactivated by TiO_2-Assisted Photocatalytic Solar Disinfection

Serial No.	Microorganism	Example	Reference
1	Bacteria	Enterobacter cloacae	Ibáñez et al. (2003)
		Escherichia coli	Sun et al. (2003), Wei et al. (1994)
		Streptococcus mutans	Saito et al. (1992)
2	Viruses	Polio virus	Watts et al. (1995)
3	Fungi	*Candida albicans*	Lonnen et al. (2004)
4	Protozoa	*Cryptosporidium parvum*	Otaki et al. (2000)

consumes hydrogen peroxide (H_2O_2). This corroborated the fact that although radicals generated at the surface of the catalyst are important initial products of the process, subsequent generation of hydrogen peroxide is likely to play a major role in the disinfection. Titanium dioxide has been found to be equally effective when used in an immobilised manner. This enhanced the prospects of employing such systems in continuous water treatment, which remain uncontaminated by particulate titanium dioxide (Salih, 2002). Immobilised photocatalyst-based systems have also shown to be effective in disinfecting water on a small scale in PET bottles. It essentially comprised a flat plastic sheet coated with TiO_2 on one (upper) surface and inserted in the PET bottle. This technology is simple and can be implemented in semiurban areas of developing countries (Duffy et al., 2004; Wegelin and De Stoop, 1999).

Further research to enhance the effectiveness of TiO_2-driven photocatalysis resulted in the revelation that metals, metal salts, and dyes improved the performance of the system. Interestingly, electric field enhancement on the catalyst surface boosted the production of hydroxyl radicals, and thereby the overall process. This could effectively inactivate the most resistant microbes, such as *Clostridium perfringens*, and their spores and cysts. This technique was called photoelectrocatalytic disinfection (Christiansen et al., 2003).

Recently, the number of additives has been utilised to enhance and accelerate the SODIS. These include chemical agents such as hydrogen peroxide, copper, and ascorbate, as well as natural food preservatives and commonly available ingredients such as lemon, lime, and vinegar. In one study 100 to 1000 mM hydrogen peroxide and 0.5 to 1% lemon and lime juice could rapidly enhance SODIS. In yet another study, stored rainwater could be very effectively disinfected and the addition of these natural ingredients could enhance the SODIS efficiency by 40% even in weak solar radiation weather conditions. This was attributed to the lowering of pH to 3. The researchers used 2.5 ml of 0.25% lemon and 1.7 ml of 0.17% vinegar (Amin and Han, 2011). Such experiments clearly point out that SODIS can be rapid and effective in both sunny and cloudy weather, leading to its potential utilisation in regions with poor sunlight.

3.4 Heat Treatment

From the Stone Age period of mankind, heating water has been practiced for over centuries now. High temperature kills most of the microbes present in water. Thus, boiling is one guaranteed way to purify water of all pathogens. It is necessary to hold the water in the boiling state for some length of time, commonly 5 or 10 min. One reason for the long period of boiling may be to inactivate bacterial spores (which can survive boiling), but these spores are unlikely to be waterborne pathogens.

Water can also be treated at below boiling temperatures, if contact time is increased. The process is similar to milk pasteurisation, and holds the water at 72°C for 15 s. Heat exchangers recover most of the energy used to warm the water. Solar pasteurisers have also been built that would heat 3 gallons of water to 65°C and hold the temperature for an hour. Regardless of the method, heat treatment does not leave any form of residual to keep the water free of pathogens in storage. Moreover, this method would be expensive for large-scale use due to the high energy requirements.

3.5 Filtration Methods

3.5.1 Reverse Osmosis

Osmosis is a phenomenon that involves the flow of small organic molecules through a semipermeable membrane from the solution side to the pure solvent side due to the concentration gradient across the membrane. At equilibrium the concentration on either side of the membrane is said to be equal and balanced by the osmotic pressure of the system. If this phenomenon is reversed, that is, pressure greater than the osmotic pressure is applied to the membrane, pure solvent will flow to the solution side through the semipermeable membrane. This is called reverse osmosis (Figure 3.3). It is applied to force water under a pressure of over 100 bars through reverse osmosis membranes that are impermeable to most contaminants. The most common use is to produce freshwater from saltwater. Larger organic molecules and all pathogens are rejected. Reverse osmosis filters are also available that use normal municipal or private water pressure to remove contaminates from

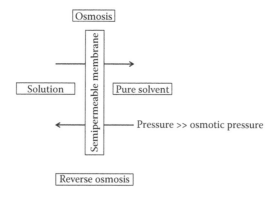

FIGURE 3.3
Phenomenon of reverse osmosis.

water. The water produced by reverse osmosis will be close to pure H_2O. Therefore, mineral intake may need to be increased to compensate for the normal mineral content of water. Reverse osmosis is effective not only in the removal of bacteria, but also in the successful removal of viruses. In a study conducted by Nederlof et al. (1998) at the Amsterdam Water Supply, more than a 3000-fold reduction of MS2 phages has been reported by using reverse osmosis membranes.

3.5.2 Microfilters

Microfilters are small-scale filters designed to remove cysts, suspended solids, protozoa, and in some cases, bacteria from water. Most filters use a ceramic or fibre element that can be periodically cleaned to restore performance as the units are continuously used. Most units, and almost all made for camping, use a hand pump to force the water through the filter. Others use gravity, either by placing the water to be filtered above the filter or by placing the filter in the water and running a siphon hose to a collection vessel located below the filter. Microfilters are effective to remove *Cryptosporidia*, but they do not remove viruses. Microfilters share a problem with charcoal filters in having bacteria grow on the filter medium. This can be overcome by impregnating the filter element with silver. A prefiltration stage to remove most of the coarse particles helps in maintaining the life of the microfilter by preventing clogging. Many microfilters may include silt prefilters, activated charcoal stages, or an iodine resin for this purpose. Allowing time for solids to settle or prefiltering also extends the filter life. Ghayeni et al. (1999) have reported that microfiltration of bacterial suspensions through 0.2 mm membranes resulted in zero culturable bacterial counts.

3.5.3 Slow Sand Filter (SSF)

Slow sand filters pass water slowly through a bed of sand. Pathogens and turbidity are removed by biological action and filtering. Typically, the filter consists of around 24 in. of sand, after which a gravel layer containing a drain pipe is embedded. The filter can be cleaned several times before the sand has to be replaced. Conventional slow sand filtration essentially consists of a box full of sand. Water enters the filter compartment above the media and flows down through the sand and, in time, will form a thin biological layer called a schmutzdecke. The combination of physical straining and biological treatment effectively removes turbidity and bacteria. *Giardia* and *Cryptosporidium* oocysts are also effectively removed from water. As the bed plugs up, the water level rises over the sand, and when it approaches an established upper level, usually the overflow, it is partially drained. The schmutzdecke is scraped off and discarded or stored for cleaning and recycling. This cycle is repeated for typically 1 to several months. Over time the sand bed depth is reduced due to repeated scrapings and new or recycled sand is refilled

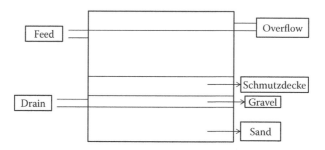

FIGURE 3.4
Slow sand filtration.

into the box. A typical slow sand filter is shown in Figure 3.4. The major advantages of slow sand filters are manifold: effective treatment for turbidity, bacteria, and cysts, no pretreatment chemicals are usually required, and generally no backwashing or automation is required. However, the major disadvantages of conventional slow sand filters are their requirement of low-turbidity raw waters and low algae counts. Turbidity can be reduced by changing the method of collection, such as building an infiltration gallery, rather than taking water directly from a creek, allowing time for the material to settle out, using a raw water tank, and prefiltering or flocculation by adding a chemical such as alum to cause the suspended material to flock together and settle out. Slow sand filters should only be used for continuous water treatment. If a continuous supply of raw water cannot be ensured, then another method should be chosen. The sand for a SSF needs to be clean and uniform, and of the correct size. The outlet of a SSF must be above the sand level and below the water level. The water must be maintained at a constant level to ensure an even flow rate throughout the filter. The flow rate can be increased by lowering the outlet pipe or increasing the water level.

When the SSF begins to work at once, optimum treatment for pathogens generally takes a week or more. During this time the water should be chlorinated to avoid contamination. After the filter stabilises, the water is safe to drink, but chlorinating of the output is still a good idea, particularly to prevent recontamination. As the flow rate slows down, the filter will have to be cleaned by draining and removing the top few inches of sand. If a geotextile filter is used, only the top ½ in. may have to be removed. As the filter is refilled, it will take a few days for the biological processes to reestablish themselves.

Slow sand filtration was initially used for wastewater treatment. Removal of microorganisms in these wastewater systems ranged between 93 and 98%. Despite the high removal efficiency of microorganisms, the final effluents showed average counts of 5.4×10^4 to 1.4×10^5 total coliforms, 5.2×10^4 to 1.3×10^5 faecal coliforms, and 1.6×10^4 to 4.7×10^4 CFU/100 ml (Botero et al., 1997). Later its application was extended to treatment of water. Schuler et al. (1991)

have used slow sand filtration for treating surface water and have reported 99.9% removal of *Giardia muris*, *Cryptosporidium*, and coliform bacteria. Esen and coworkers have studied the efficiency of slow sand filtration in removing algae from high-rate pond effluents. When agricultural sandy soil with an effective grain size of 0.08 mm was used as the filter medium, an average filtration rate of about 1.3 m^3/m^2 day was obtained. If filtration was preceded by sedimentation, the duration of a filtration run was about 100 h. At the end of each run, the filter was cleaned by scraping off the top few centimeters of the filter bed. The filtered effluent consistently had a biochemical oxygen demand (BOD) value less than 20 mg/l and undetectable faecal coliforms. The filter media, being rich in organic matter and having moisture-retaining properties, can be used as a soil conditioner (Esen et al., 1991).

Ellis has investigated slow sand filtration as a means of tertiary treatment for secondary effluents derived from conventional aerobic, biological treatment processes operating with municipal wastewaters. The basic slow sand filtration unit used consisted of a 140 mm internal diameter perspex cylinder, 2.65 m in height, containing a 950 mm depth of fine sand. Treatment rates were either 3.5 or 7.0 m/day (m^3/m^2/day = m/day) and the sand used was of an effective size, initially of 0.3 mm and then later of 0.6 mm. This investigation has demonstrated that a laboratory-scale slow sand filtration unit is capable of consistently removing at least 90% of the suspended solids, more than 65% of the remaining BOD, and over 95% of the coliform organisms from the settled effluent from an operational percolating filter plant. The length of operational run averaged 20 days at 3.5 m/day and 13 days at 7.0 m/day (Ellis, 1987). In a very recent study, SSF has been demonstrated to be a promising alternative for large-scale river bank filtration (RBF). It has shown to be very effective when sources of water are faecally polluted (Bauer et al., 2011).

Turbidity has been shown to be responsible for the clumping of microbes in water, thereby reducing the effectiveness of water disinfection. In a very interesting study, a low-tech option for water treatment using a bicycle-powered filtration system made of sand and ceramic effectively demonstrated the reduction of not only the turbidity levels (98%), but also total coliforms (67%) and faecal coliforms (89%) (McBean, 2008).

3.5.4 Activated Charcoal Filter

Activated charcoal filters water and removes impurities by the adsorption principle. Therefore, most of the chemical and biological impurities get adsorbed onto the surface of activated charcoal. Over a period of time the adsorptive capability of the activated charcoal may diminish and the adsorbed microbes may even start colonising due to the favourable environment provided by the charcoal particles. Water passing through such a filter will result in the addition of pathogens rather than the removal of them. In order to prevent this, silver impregnation of the charcoal particles is resorted to. Alternatively, activated charcoal in combination with a chemical disinfectant such as chlorine or iodine may be employed wherein the chemicals kill the pathogens and the activated charcoal removes the treatment chemicals. The bed of carbon must be deep enough for adequate contact with the water. Production designs use granulated activated charcoal (effective size of 0.6 to 0.9 mm for maximum flow rate). Powered charcoal can also be mixed with water and filtered out later.

Activated carbon at a dose of 100 mg/dm^3 has been used for treating surface water in Poland (Konieczny and Klomfas, 2002). To obtain maximum efficiency of the activated carbon, in the adsorption process, it is desirable to have the greatest possible surface area in the smallest practical volume. The use of special manufacturing techniques results in highly porous charcoals that have surface areas of 300 to 2000 m^2/g. The huge surface area of activated charcoal gives it countless bonding sites. When certain bacteria pass next to the carbon surface, they attach to the surface and are trapped. This results in high bacterial removal efficiency and a greater system economy. Granulated activated carbon (GAC) adsorption filters are also used routinely for drinking water treatment. They have been reported to effectively remove water microbes, including MS2, *E. coli*, and spores of *Cryptosporidium parvum* and *Giardia lamblia*, largely by the process of attachment (Hijnen et al., 2010).

Extensive work in the area of filtration has led to the investigation of different types of membranes for the effective removal of waterborne pathogens. Several synthetic membranes have been experimented with, including nanocomposite ones. In a study related to the removal of bacteria and viruses from water using layered double hydroxide (LDH) nanocomposites made of zinc-magnesium or magnesium-aluminium, it was observed that the indicator viruses (MS2 and φX174) and E. coli (indicator bacteria) from synthetic groundwater could be removed by the process of adsorption with an efficiency of 99%. The LDH was operated either in a suspension or as a packed column. The column was also tested for the removal of heterotropic bacteria from raw river water in both modes. While removal efficiencies were still high (87–99%), the adsorption capacities of the two kinds of LDH were four to five orders of magnitude lower than when exposed to synthetic groundwater (Jin et al., 2007).

Nonavailability of suitable materials for slow sand filtration or any other filtration techniques mentioned above may necessitate the incorporation of alternative mechanical filtration methods as a pretreatment to chemical or physical disinfection. Some municipal water treatments may also use rapid sand filtration where water is allowed to pass through a bed of sand. Physical removal of waterborne *Crytosporidium* oocysts and *Giardia* cysts is ultimately achieved by properly functioning conventional filters, provided that effective pretreatment of the water is applied. Disinfection by chemical or physical methods is finally required to inactivate the infectious life stages of these organisms (Betancourt and Rose, 2004). Pretreatment with a coarse filtration is generally used with chlorination and UV radiation to reduce turbidity and maintain high effectiveness. Cysts and worm eggs are resistant to chlorination and UV but can be filtered out relatively easily. Slow sand filtration is lowest in cost but requires high investment in labor. Every technique has its own merits and demerits and can be selected based on myriad factors, such as type of source water, type of microbes being targeted, cost, and economics, to name a few.

3.6 Distillation

The process of evaporation and condensation of water is called distillation. It results in water that is extremely pure and potable. Equipment used to carry our distillation is called distillers. It comprises a heating coil that boils water to generate steam, which inactivates bacteria, viruses, and protozoan cysts. The generated steam then rises and passes to another section, which contains the condensation coil. Here, the steam condenses to generate distilled water, which is stored in a separate container. Water remaining in the boiling chamber has a much higher concentration of impurities, such as inorganic compounds and large nonvolatile organic compounds that are not vapourised. The unit is provided with a drain to discard these impurities.

Volatile organic compounds (VOCs) are those compounds that have their boiling point very close or below that of water. Therefore, they are likely to vapourise early in the distillation process and, if not removed, VOCs condense back to the liquid along with water, making it impure. There are several strategies that are routinely adopted to combat this issue. Provisions of gas vents to allow the escape of VOCs before they enter the cooling section, a fractional distiller, where the VOCs are cooled and condensed in a separate section, or use of an activated carbon filter to remove VOCs from condensed water are some techniques to solve the problem of water contamination by VOCs.

Equipment employed for distillation is called distillers, which are constructed using materials such as stainless steel, aluminium, or plastic. These materials can be easily sanitised and kept contaminant-free, which also allows storage of the distilled water in them without concern of recontamination. Distillers can be operated either in batch or continuous mode, depending on the demand for potable water. For most household purposes, the batch type would suffice, which allows water to be poured directly into it. The heating element, usually a coil provided, and the unit shut off when all the water in the chamber evaporates. The capacity of batch distillers generally ranges from 1 to 10 gallons. The continuous mode is usually a floor unit that produces more than 10 gallons per day and is connected to a water supply line. The distilled water is stored in a container, and as it is utilised, the unit automatically starts producing distilled water to replenish the used amount. In this manner the continuity is maintained. The water and impurities remaining in the boiling chamber are periodically removed through a discharge line, termed *bleed* or *blowdown*.

The most significant operational cost for distillation is the electricity required to heat and vapourise the water to generate steam (other costs include cleaning solution and, if equipped, AC filter replacement). Operational costs are directly dependent on the amount of distilled water used daily (which determines how often the unit operates). Due to a very high latent heat of vapourisation of water, a large amount of energy requirement makes it an expensive alternative to physical disinfection. Moreover, it is likely that contaminants with boiling points below that of water are likely to condense along with water, leading to contamination. Thus, distillation removes both pathogens and dissolved minerals and salts from water. This process is strongly dependent on a heating source that could be solar radiation (used in solar stills) or simple heating and cooling coils. More efficient distillation plants use multiple-effect evaporation and a vapour compression cycle where the water is boiled off at atmospheric pressure, the steam is compressed, and the condenser condenses the steam above the boiling point of the water in the boiler, returning the heat of condensation to the boiling water. The hot condensed water is run through a second heat exchanger that heats up the water feeding into the boiler. This is the method used in most commercial plants because it is more economical.

Distillation removes a spectrum of contaminants from water, such as sodium, calcium, magnesium (hardness compounds), dissolved salts, and microbes such as bacteria, viruses, and protozoan cysts. Some amount of oxygen is also removed, along with the trace metals, imparting a bland or flat taste to the water, which the consumer may not enjoy. Even though most of the impurities are removed by distillation, trace amounts of the original contaminants may still be present in the distilled water, especially when the impurities are volatile and their boiling points are similar or just below that of water. Certain pesticides, volatile solvents, and volatile organic compounds such as benzene and toluene, with boiling points close to or below that of water, vapourise along with water as it is boiled. Generally an additional treatment, like the carbon bed adsorption, may be employed just before condensation in order to eliminate such compounds. Microorganisms are generally inactivated by the boiling process. However, if the distiller is idle for an extended period, bacteria can be reintroduced from the outlet spout of the distiller and may recontaminate the water.

Thus, drinking water treatment using distillation is one option for treating water problems. Operated properly, distillation can remove up to 99.5% of impurities from water, including bacteria, metals, nitrate, and dissolved solids. Operation costs for distillation can be among the highest for home drinking water treatment systems. Selection of a distillation unit should be based on water analysis and assessment of the individual homeowner's needs and situation. Regular maintenance of the unit is a critical factor in maintaining its effectiveness.

3.7 Electrochemical Oxidation

Electrochemical oxidation does not involve the direct addition of chemicals into the water to be treated. However, the electrodes introduced into the system generate chemical species that are generally responsible for disinfection. The technique is very simple, involving the use of two electrodes, an anode and a cathode, and the application of direct current (DC) current to the system. This brings about the generation of the disinfectant species for treating water. It is interesting to note that this simple technique, although reported to have been utilised as early as the nineteenth century, has not really gained much importance subsequently. Sporadic research in this arena has occurred over the last few decades owing to some of the advantages of electrochemical oxidation over the conventional methods, such as chlorination and other chemical methods. The advantages are mainly no transport, storage, and dosage of disinfectants required. It is also important to note that the use of electrochemical oxidation is possible some distance away from the electric supply grid, which is very advantageous for remote rural areas not accessible to electricity. The upcoming importance of hybrid methodologies of water disinfection owing to their better efficiency and economics with the added benefit of lower disinfection by-products points to the possibility of using electrochemical oxidation in conjunction with other techniques. Not much has been reported on this, though. The following paragraph describes briefly the fundamental mechanism of the process.

It is well established that when an anode and a cathode are dipped in water, oxygen is produced at the anode along with liberation of hydrogen, and hence the slight acidification near the vicinity of the anode. Hydrogen is formed at the cathode. Free chlorine is also produced from the chloride content of water, which acts as the disinfection species. It basically involves the hydrolysis of chlorine in water to produce hypochlorous acid, which in turn produces the hypochlorite ions. The sum of the hypochlorous acid and the hypochlorite ion together is called free chlorine, and the disinfection action of free chlorine is due to atomic oxygen release. Ozone that is routinely used for water disinfection can also be produced electrochemically. However, for the former application, it is generally generated by the corona discharge method (discussed in the previous section). Diamond solid polymer electrolyte (SPE) sandwich electrodes are used to generate ozone electrochemically for water disinfection. Hydrogen peroxide, which is known for its water disinfecting properties, can be produced at the cathode and has been reported in the literature (Dhar et al., 1981; Drogui et al., 2001). Owing to its lower oxidation potential, hydrogen peroxide is generally less effective than chlorine ozone. Over a period of time, water treatment reactors based on electrochemical disinfection have been developed, and research papers focussing on these aspects can be found in literature. For instance, Matsunaga et al. (1992) have reported a novel electrochemical oxidation reactor based on carbon cloth electrode for inactivation of *E. coli*. The disinfection rate reported was 6×10^2 cells/cm^3/h at a cell concentration of 10^2 cells/cm^3. Continuous sterilisation was carried out for a period of 24 h. The mode of inactivation reported is due to electrochemical reaction between electrode and microbial cells, which is supposed to be mediated by intracellular coenzyme A. This was the first demonstration of effective disinfection by electrochemical oxidation (Matsunaga et al., 1992).

3.8 Water Disinfection by Microwave Heating

Microwave heating involves the use of microwave radiation (2.25 GHz) to inactivate microbes. Over the last decade its use has risen for various applications, including medical waste sterilisation, food disinfection, and so on. Microwave heating has several advantages over conventional radiant heating, such as rapid heating rates and uniform heating, resulting in less energy requirements (Lidstrom et al., 2001). Several studies are reported for the inactivation of microorganisms, especially bacteria by microwave irradiation, and the mechanism has also been debated. Although the mechanism of action of microwave heating on bacteria and viruses, especially on the DNA, was studied, the exact mode is yet to be unravelled. This is because it is still believed that microwave heating results in two phenomena upon irradiation: thermal and nonthermal effect. The former is essentially brought about due to the induction of collisional deactivation owing to Brownian movement, resulting in thermal heating (Larhed et al., 2002). Nonthermal effects include those that are not associated with a rise in temperature, which occur when the microwave fields cause a polarised side chain of macromolecules to line up with the direction of the electric field. This leads to the breakage of hydrogen bonds, resulting in the denaturation and coagulation of molecules. Thus, microwave-induced structural and orientation effects within the irradiation medium are called the nonthermal effect. Interestingly, the exact mechanism of microwave-induced inactivation of microorganisms is not yet clearly understood.

Many researchers have compared microwave heating and conventional radiant heating, and contradictory results are reported in literature. On one hand, there are reports stating no differences in the deactivation rates between the two processes, when investigated for different microorganisms, such as *Bacillus subtilis* (Jeng et al., 1987), *E. coli* (Fujikawa et al., 1992), and *Clostridium* species (Welt and Tong, 1993), while on the other hand, some researchers report that microwave heating, especially using pulsed microwave, results in higher deactivation rates than conventional radiant heating, and the study also revealed that nonthermal effects were also responsible for inactivation (Shin and Pyun, 1997). Initially, many researchers believed that the inactivation of microbes by microwave was only due to dielectric heating (Goldblith and Wang, 1967). Inactivation by microwave was attributed to several reasons, including alteration of biochemical processes in bacteria such as *E. coli* and *B. subtilis* (Kazbekov and Vyacheslavov, 1987). However, this was challenged by some who believed that the nonthermal effects could be due to the lack of accurate measurements of the time-temperature history (Heddleson et al., 1994).

Over a period of time, several studies were undertaken to ascertain the effect of factors such as ionic strength and the presence of metals on microwave disinfection. It is interesting to note that an almost 10-fold increase in reduction of *E. coli* was achieved when microwave irradiation was carried out in the presence of molar concentrations of NaCl and KCl (Watanabe et al., 2000). Transitional metals and their role in bacteria have been known for a long time (Beveridge and Doyle, 1989). Several transitional metals, such as calcium, potassium, magnesium, zinc, and so forth, are required for the normal functioning of the bacteria, such as in osmotic regulation, biochemical processes, and as micronutrients. On the contrary, some transitional metals like mercury, silver, aluminium, and lead do not have any biological role and are supposed to be toxic to bacteria. The toxicity also depends on the type of metal and its concentration. Many such metals have been employed in combination with established physical and chemical techniques already discussed in this book. Similarly, attempts were also made to combine metals with microwave irradiation for bacterial inactivation. In one such work, the effect of iron and cobalt ions on microwave inactivation of *E. faecalis, S. aureus*, and *E. coli* was studied. It was observed that there was a rapid reduction in the viable bacteria during microwave treatment in the presence of metal ions. This was attributed to the increased metal ion penetration into the microorganism (Benjamin et al., 2007).

3.9 Conclusion

Physical disinfection techniques appear to be a promising alternative to the established chemical disinfection methods. One of the major limitations of chemical disinfection, that is, potentially carcinogenic disinfection by-product formation, can be alleviated to a large extent by following physical methods. Moreover, the routine taste and odour problem as well as the emergence of chemical-resistant microorganisms may be an added reason to explore the use of such alternatives. However, the preceding discussion indicates that not all physical techniques are devoid of DBP formation, UV disinfection being one such method. This leads one to contemplate both the merits and demerits of any technique, and also paves the way for deciding the best option for a given scenario. On the positive side, physical methods such as UV irradiation and membrane systems are sufficiently developed to be considered alternative disinfection strategies, and in addition to their excellent disinfection ability, they have benefits such as improved chemical safety and customer aesthetics. Disadvantages include difficulty in monitoring process performance, lack of residual disinfection in the distribution systems, higher energy consumption, and higher costs. Costs increase with the degree of particle size removal, that is, microfiltration < ultrafiltration < nanofiltration < reverse osmosis, and the flux/m^2 is independent of the scale. Generation of liquid waste containing impurities are concentrated during the process. This may also require some waste treatment to detoxify the chemical compounds and kill the microorganisms that may be accumulated on the membranes over a period of time. The additional cost for this kind of waste disposal may be overlooked in the overall estimation of the physical treatment technique. The main implication for water treatment and process monitoring of changing to alternatives of chemical disinfection would be the need to ensure the efficacy of disinfection. It appears that a low dose of an appropriate chemical disinfectant needs to be combined with these physical techniques to ensure disinfection efficiency, especially posttreatment, to prevent regrowth in the distribution systems. Finally, the application of any disinfection method has to be done on a case-to-case basis. This necessitates taking into account several important parameters, such as the type of water to be disinfected, treatment method being used, and scale of operation. Since no single method is the best alternative for water disinfection, consideration of hybrid techniques appears to be the need of the hour.

Questions

1. Explain the need for physical disinfection techniques for water disinfection.

2. Describe the classification of UV radiation and state which range is effective as a disinfectant.

3. How is UV light generated?

4. With the help of a neat diagram explain the microscreening/UV method for inactivation of *Giardia* cysts and *Cryptosporidium* oocysts.

5. Explain the concept of pulsed UV for water disinfection.

6. Describe the mechanism of action of UV rays.

7. Explain the phenomena of photoreactivation and its implication in water disinfection.

8. What is dark repair? What are the latest findings related to it?

9. How are bacteria, viruses, and protozoa inactivated by the UV technique?
10. What is the relationship between UV dose and coliform survival?
11. Describe the various factors affecting UV inactivation of microorganisms.
12. How does UV reactor geometry affect water disinfection?
13. What are the effects of microbial clumping and turbidity on the UV disinfection efficiency?
14. Discuss DBPs during UV treatment of water.
15. With a neat labeled diagram, describe the mechanism of action of solar disinfection.
16. Discuss the factors affecting SODIS.
17. How does light intensity and dissolved oxygen affect the SODIS process?
18. Does the type of microorganism to be treated have an impact on the overall efficiency of SODIS?
19. Write a note on TiO_2-assisted photocatalytic SODIS.
20. Discuss the recent developments in SODIS.
21. Describe heat as a method of water disinfection. What are its pros and cons?
22. Describe reverse osmosis for water disinfection.
23. Explain microfiltration as a technique of water disinfection.
24. Explain the principle of slow sand filtration with a neat labeled diagram.
25. What is the role of an activated carbon filter in water disinfection?
26. Explain the principle and working of the distillation process for water disinfection.
27. Discuss the overall merits and demerits of physical disinfection methods and the challenges ahead.

References

Acra, A., and Ayoub, G. (1997). Experimental evalution of a novel photodynamic water disinfection technique. *J. Water Sci. Res. Technol.* 46:218–223.

Acra, A., Jurdi, M., Múallem, Y., Karahagopian, Y., and Raffoul, Z. (1990). *Water disinfection by solar radiation.* International Disinfection by Solar Radiation. International Development Research Center, Ottowa, Canada.

Acra, A., Raffoul, Z., and Karahagopian, Y. (1984). *Solar disinfection of drinking water and oral rehydration solutions: Guidelines for household application in developing countries.* New York: UNICEF.

Amin, M.T., and Han, M.Y. (2011). Improvement of solar base drain water disinfection by using lemon and vinegar as catalysts. *Desalination* 276(1–3):416–424.

Anon. (2001). World Water Day 2001 Web site. Available at http://www.worldwater-day.org.reportch4html (accessed September 26, 2003).

Anon. (2002). World Health Report 2002. Reducing risks, promoting healthy life. *Aqua* 41(2):95.

AWWA and ASCE (American Society of Civil Engineers). (1990). *Water treatment plant design,* 2nd ed. New York: McGraw-Hill.

Banerjea, R. (1950). The use of potassium permanganate in the disinfection of water. *Ind. Med. Gaz.* 85:214–219.

Bauer, R., Dizer, H., Graeber, I., Rosenwinkel, K.H., and López-Pila, J.M. (2011). Removal of bacterial fecal indicators, coliphages and enteric adenoviruses from waters with high fecal pollution by slow sand filtration. *Water Res.* 45(2):439–452.

Benjamin, E. III, Reznik, A., Benjamin, E., and Williams, A.L. (2007). Mathematical models of cobalt and iron ions catalyzed microwave bacterial deactivation. *Int. J. Environ. Res. Public Health* 4(3):203–210.

Betancourt, W.Q., and Rose, J.B. (2004). Drinking water treatment processes for removal of *Cryptosporidium* and *Giardia*. *Vet. Parasitol.* 126(1–2):219–234.

Beveridge, T.J., and Doyle, R.J. (1989). *Metal ions and bacteria.* New York: Wiley.

Bocharov, D.A. (1970). Content of nucleic acids in bacterial suspension after influence on it of chlorinated and alkaline solutions. Trudy Vsesoiuznvi Nauchno-issledovatel'skii Institute, *Veterinarnoi Sanitarii* (Soviet Union), 34:242–252.

Botero, L., Montiel, M., Estrada, P., Villalobos, M., and Herrera, L. (1997). Microorganism removal in wastewater stabilisation ponds in Maracaibo, Venezuela. *Water Sci. Technol.* 35(11–12):205–209.

Brion, G.M., and Silverstein, J. (1999). Iodine disinfection of a model bacteriophage, MS2, demonstrating apparent rebound. *Water Res.* 33(1):169–179.

Buth, J.M., Arnold, W.A., and McNeil, K. (2007). Unexpected products and reaction mechanisms of the aqueous chlorination of cimetidine. *Environ. Sci. Technol.* 41(17):6228–6233.

Campbell, A.T., Robertson, L.J., Snowball, M.S., and Smith, H.V. (1995). Inactivation of oocysts of *Cryptosporidium parvum* by ultraviolet. *Water Waste Dig.* 29(11):2583–2586.

Canosa, P., Rodriguez, I., Rubi, E., Negreira, N., and Cela, R. (2006). Formation of halogenated by-products of parabens in chlorinated water. *Anal. Chim. Acta* 575:106–113.

Cantwell, R.E., and Hofmann, R. (2008). Inactivation of indigenous coliform bacteria in unfiltered surface water by ultraviolet light. *Water Res.* 42(10–11):2729–2735.

Carlson, D.A. (1985). *Project summary: Ultraviolet disinfection of water for small water. supplies.* Office of Research and Development, Cincinnati, OH: EPA/600/S2-85/092.

Chilvers, K.F., Reed, R.H., and Perry, J.D. (1999). Phototoxicity of rose bengal in mycological media— Implications for laboratory practice. *Lett. Appl. Microbiol.* 28:103–107.

Choi, Y., and Choi, Y.-J. (2010). The effects of UV disinfection on drinking water quality in distribution systems. *Water Res.* 44(1):115–122.

Christiansen, P.A., Curtis, T.P., Egerton, T.A., Kosa, S.A.M., and Tinlin, J.R. (2003). Photoelectrochemical and photocatalytic disinfection of *E. coli* suspensions by titanium dioxide. *Appl. Catal. B Environ.* 41:371–386.

Ciochetti, D., and Metcalfe, R. (1984). Pasteurization of naturally contaminated water with solar energy. *Appl. Environ. Microbiol.* 47:223–228.

Clancy, J.L., Hargy, T.M., Marshall, M.M., and Dyksen, J.E. (1998). UV light inactivation of *Cryptosporidium* oocysts. *J. Am. Water Works Assoc.* 90(9):92–102.

Combs, R., and McGuire, P. (1989). Back to basics—The use of ultraviolet light for microbial control. *Ultrapure Water J.* 6(4):62–68.

Conroy, R.M., Elmore-Meegan, M., Joyce, T., McGuigan, K.G., and Barnes, J. (1996). Solar disinfection of drinking water and diarrhoea in Maasai children: A controlled field trial. *Lancet* 348:1695–1697.

Conroy, R.M., Elmore-Meegan, M., Joyce, T., McGuigan, K., and Barnes, J. (1999). Solar disinfection of water reduces diarrhoeal disease: An update. *Arch. Dis. Child.* 81:337–338.

Conroy, R.M., Meegan, M.E., Joyce, T., McGuigan, K., and Barnes, J. (2001). Solar disinfection of drinking water protects against cholera in children under 6 years of age. *Arch. Dis. Child.* 85:293–295.

Curtis, T.P., Mara, D.D., and Silva, S.A. (1992). Influence of pH, oxygen and humic substances on ability of sunlight to damage fecal coliforms in waste stabilization pond water. *Appl. Environ. Microbiol.* 58:1335–1343.

Davies, C.M., and Evison, L.M. (1991). Sunlight and the survival of enteric bacteria in natural waters. *J. Appl. Bacteriol.* 70:265–274.

Davies-Colley, R.J., Donnison, A.M., and Speed, D.J. (1997). Sunlight wavelengths inactivating faecal indicator microorganisms in waste stabilisation ponds. *Water Sci. Technol.* 35:219–225.

DeMers, L.D., and Renner, R.C. (1992). *Alternative disinfection technologies for small drinking water systems.* Denver, CO: AWWA and AWWARF.

Dhar, H.P., Bockris, J.O., and Lewis, D.H. (1981). Electrochemical inactivation of marine bacteria. *J. Electrochem. Soc.* 128(1):229–231.

Downes, A., and Blunt, T.P. (1887). Researches on the effect of light upon bacteria and other organisms. *Proc. Roy. Soc. (London)* 26:488–500.

Drogui, P., Elmaleh, S., Rumeau, M., Bernard, C., and Rambaud, A. (2001). Hydrogen peroxide production by water electrolysis: Application to disinfection. *J. Appl. Electrochem.* 31(8):877.

Duffy, E.F., Al-Touati, F., Kehoe, S.C., McLoughlin, O.A., Gill, L., Gernjak, W., Oller, I., Maldonado, M.I., Malato, S., Cassidy, J., Reed, R.H., and McGuigan, K.G. (2004). A novel TiO_2-assisted solar photocatalytic batch-process disinfection reactor for the treatment of biological and chemical contaminants in domestic drinking water in developing countries. *Solar Energy.* 77:649–655.

Ellis, C., and Wells, A.A. (1941). *The chemical action of ultraviolet rays.* New York: Reinhold Publishing Co.

Ellis, K.V. (1987). Slow sand filtration as a technique for the tertiary treatment of municipal sewages. *Water Res.* 21(4):403–410.

Ellis, K.V., and Van Vree, H.B.R.J. (1989). Iodine used as water disinfectant in turbid water. *Water Res.* 23(6):671–676.

Slade, J.S., Harris, N.R., and Chisholm, R.G. (1986). Disinfection of chlorine-resistant enteroviruses in groundwater by ultraviolet radiation. *Water Sci. Technol.* 189(10):115–123.

Esen, I.I., Puskas, K., Banat, I.M., and Al-Daher, R. (1991). Algae removal by sand filtration and reuse of filter material. *Waste Manage.* 11(1–2):59–65.

Fujikawa, H., Ushioda, H., and Kudo, Y. (1992). Kinetics of *Escherichia coli* destruction by microwave irradiation. *Appl. Environ. Microbiol.* 58(3):920–924.

Fraker, L.D., Gentile, D.A., Krivoy, D., Condon, M., and Backer, H.D. (1992). *Giardia* cyst inactivation by iodine. *J. Wilderness Med.* (3)4:351–357.

Gall, R.J. (1978). Chlorine dioxide, an overview of its preparation, properties and uses. In R.G. Rice and J.A. Cotruvo (eds.), *Ozone/chlorine dioxide. Oxidation product of organic materials.* Cleveland, OH: Ozone Press Int., pp. 356–382.

Gerba, C.P., Johnson, D.C., and Hasan, M.N. (1997). Efficacy of iodine water purification tablets against *Cryptosporidium* oocysts and *Giardia* cysts. *Wilderness Environ. Med.* 8(2):96–100.

Ghayeni, S.B.S., Beatson, P.J., Fane, A.J., and Schneider, R.P. (1999). Bacterial passage through microfiltration membranes in wastewater applications. *J. Membrane Sci.* 153(1):71–82.

Goldblith, S.A., and Wang, D.I. (1967). Effect of microwaves on *Escherichia coli* and *Bacillus subtilis*. *Appl. Microbiol.* 15(6):1371–1375.

Gómez-Couso, H., Fontán-Sainz, M., McGuigan, K.G., and Ares-Mazás, E. (2009). Effect of the radiation intensity, water turbidity and exposure time on the survival of *Cryptosporidium* during simulated solar disinfection of drinking water. *Acta Tropica* 112(1):43–48.

Goosen, N., and Moolenaar, G.F. (2008). Repair of UV damage in bacteria. *DNA Repair* 7(3):353–379.

Guillard, J.F., Derrien, A., Gourmelon, M., and Pommepuy, M. (1997). T90 as a tool for engineers: Interests and limits. *Water Sci. Technol.* 35:277–281.

Harper, J.C., Christensen, P.A., Egerton, T.A., Curtis, T.P., and Guzlazuardi, J. (2001). Effect of catalyst type on the kinetics of the photoelectrochemical disinfection of water inoculated with *E. coli*. *J. Appl. Electrochem.* 31:623–628.

Hazen and Sawyer. Environmental Engineers & Scientists. (1992). *Disinfection alternatives for safe drinking water*. New York: Van Nostrand Reinhold.

Heaselgrave, W., and Kilvington, S. (2011). The efficacy of simulated solar disinfection (SODIS) against *Ascaris, Giardia, Acanthamoeba, Naegleria, Entamoeba,* and *Cryptosporidium*. *Acta Tropica* 119(2–3):138–143.

Heddleson, R.A., Doores, S., and Anantheswaran, R.C. (1994). Parameters affecting destruction of Salmonella spp. by microwave heating. *J. Food Sci.* 59(2):447–451.

Hijnen, W.A.M., Suylen, G.M.H., Bahlman, J.A., Brouwer-Hanzens, A., and Medema, G.J. (2010). GAC adsorption filters as barriers for viruses, bacteria and protozoan (oo)cysts in water treatment. *Water Res.* 44(4):1224–1234.

Hsu, Y. (1964). Resistance of infectious RNA and transforming DNA to iodine which inactivates *f*2 phage and cells. *Nature* 203:152–153.

Hsu, Y.C., Nomura S., and Kruse, C.W. (1966). Some bactericidal and virucidal properties of iodine not affecting infectious RNA and DNA. *Am. J. Epidemiol.* 82:317–328.

Hughes, W.L. (1957). The chemistry of iodination. *Ann. N.Y. Acad. Sci.* 70:3–18.

Ibáñez, J.A., Litter, M.I., and Pizarro, R.A. (2003). Photocatalytic bactericidal effect of TiO$_2$ on *Enterobacter cloacae*. Comparative study with other gram (−) bacteria. *J. Photochem. Photobiol. A Chem.* 157:81–85.

Jagger, J. (1967). *Introduction to research in ultraviolet photobiology.* Englewood Cliffs, NJ: Prentice-Hall.

Jeffrey, W.H., Aas, P., Lyons, M.M., Coffin, R.B., Pledger, R.J., and Mitchell, D.L. (1996). Ambient solar radiation-induced photodamage in marine bacterioplankton. *Photochem. Photobiol.* 64:419–427.

Jeng, D.K., Kaczmarek, K.A., Woodworth, A.G., and Balasky, G. (1987). Mechanism of microwave sterilization in the dry state. *Appl. Environ. Microbiol.* 53(9):2133–2137.

Jin, S., Paul, H., Fallgren, J.M., and Morris, Q.C. (2007). Removal of bacteria and viruses from waters using layered double hydroxide nanocomposites. *Sci. Technol. Adv. Mater.* 8(1–2):67–70.

Johnson, R.C. (1997). Getting the jump on *Cryptosporidium* with UV. *Opflow* 23(10):1.

Joyce, T.M., McGuigan, K.G., Elmore-Meegan, M., and Conroy, R.M. (1996). Inactivation of fecal bacteria in drinking water by solar heating. *Appl. Environ. Microbiol.* 62:399–402.

Jungfer, C., Schwartz, T., and Obst, U. (2007). UV-induced dark repair mechanisms in bacteria associated with drinking water. *Water Res.* 41(1):188–196.

Kai, H.O.P., Tanski, H.H., and Hänninen, O.O.P. (2003). Accumulation of silver from drinking water into cerebellum and musculus soleus in mice. *Toxicology* 186(1-2):151–157.

Karanis, P., Maier W.A., Seitz, H.M., and Schoenen D. (1992). UV sensitivity of protozoan parasites. *J. Water Supply Res. Technol. Aqua* 41(2):95–100.

Kazbekov, E.N., and Vyacheslavov, L.G. (1987). Effects of microwave irradiation on some membrane-related processes in bacteria. *Gen. Physiol. Biophys.* 6(1):57–64.

Kehoe, S.C., Joyce, T.M., Ibrahim, P., Gillespie, J.B., Shahar, R.A., and McGuigan, K.G. (2001). Effect of agitation, turbidity, aluminium foil reflectors and container volume on the inactivation efficiency of batch-process solar disinfectors. *Water Res.* 35:1061–1065.

Konieczny, K., and Klomfas, G. (2002). Using activated carbon to improve natural water treatment by porous membranes. *Desalination* 147(1–3):109–116.

Krasner, S.W., Weinberg, H.S., Richardson, S.D., Pastor, S.J., Chinn, R., Sclimenti, M.J., Onstad, G.D., and Thruston, A.D. (2006). Occurrence of a new generation of disinfection by-products. *J. Environ. Sci. Technol.* 40(23):7175–7185.

Larhed, M., Moberg, C., and Hallberg, A. (2002). Microwave-accelerated homogeneous catalysis in organic chemistry. *Acc. Chem. Res.* 35:717–727.

Lee, Y., and Nam, S. (2002). Reflection on kinetic models to the chlorine disinfection for drinking water production. *J. Microbiol.* 40(2):119–124.

Lidstrom, P., Tierney, J., Wathey, B., and Westman, J. (2001). Microwave assisted organic synthesis—A review. *Tetrahedron* 57:9225–9283.

Lonnen, J., Kilvington, S., Kehoe, S.C., Al-Touati, F., and McGuigan, K.G. (2005). Solar photocatalytic disinfection of protozoan, fungal and bacterial microbes in drinking water. *Water Res.* 39(5):877–883.

Malley Jr., J.P., Shaw, J.P., and Ropp, J.R. (1995). *Evaluations of byproducts by treatment of groundwaters with ultraviolet irradiation.* Denver, CO: AWWARF and AWWA.

Masschelein W.J. (1979). *Chlorine dioxide: Chemistry and environmental impact of oxychlorine compounds.* Ann Arbor, MI: Ann Arbor Science.

Matsunaga, T., Nakasono, S., Takamuku, T., Burgess, J.G., Nakamura, N., and Sode, K. (1992). Disinfection of drinking water by using a novel electrochemical reactor employing carbon-cloth electrodes. *Appl. Environ. Microbiol.* 58(2):686–689.

Matsunaga, T., Tomoda, R., Nakajima, T., and Wake, H. (1985). Photo-electrochemical sterilisation of microbial cells by semiconductor powders. *FEMS Microbiol. Lett.* 29:211–214.

McBean, E.A. (2008). Evaluation of a bicycle-powered filtration system for removing "clumped" coliform bacteria as a low-tech option for water treatment. Presented at the Water and Sanitation in International Development and Disaster Relief (WSIDDR) International Workshop, Edinburgh, Scotland, May 28–30.

McGuigan, K.G., Joyce, T.M., Conroy, R.M., Gillespie, J.B., and Elmore-Meegan, M. (1998). Solar disinfection of drinking water contained in transparent plastic bottles: Characterizing the bacterial inactivation process. *J. Appl. Microbiol.* 84:1138–1148.

Méndez-Hermida, F., Ares-Mazás, E., McGuigan, K.G., Boyle, M., Sichel, C., and Fernández-Ibáñez, P. (2007). Disinfection of drinking water contaminated with *Cryptosporidium parvum* oocysts under natural sunlight and using the photocatalyst TiO2. *J. Photochem. Photobiol. B Biol.* 88(2–3):105–111.

Murov, S.L. (1973). *Handbook of photochemistry.* New York: Marcel Dekker.

NAS. (1987). *Drinking water and health: Disinfectants and disinfectant by-products, Vol. 7.* Washington, DC: National Academy Press. Southern Research Institute, Birmingham, AL. 1988.

Nederlof, M.M., Kxuithof, J.C., Herman, J., Koning, M., van der Hoek, J.-P., Bonn, P.A.C. (1998). Integrated multi-objective membrane systems application of reverse osmosis at the Amsterdam Water Supply. *Desalination* 119:263–273.

No, H.K., Park, N.Y., Lee, S.H., and Meyers, S.P. (2002). Antibacterial activity of chitosans and chitosan oligomers with different molecular weights. *Int. J. Food Microbiol.* 74:65–72.

Oates, P.M., Shanahan, P., and Polz, M.F. (2003). Solar disinfection (SOLDIS): Simulation of solar radiation for global assessment and application for point-of-use water treatment in Haiti. *Water Res.* 37:47–54.

Ollis, E., Pelizzetti, E., and Serpone, N. (1991). Destruction of water contaminants. *Environ. Sci. Technol.* 25:1523–1529.

Otaki, M., Hirata, T., and Ohgaki, S. (2000). Aqueous microorganisms inactivation by photocatalytic reaction. *Water Sci. Technol.* 42:103–108.

Prat, R., Nofre, C., and Cier, A. (1968). Effects of sodium hypochlorite, ozone and ionizing radiation on the pyrimidine constituents of *Escherichia coli* (in French). *Annales de L'Institut Pasteur* 114:594–607.

Reed, R.H., (1997a). Sunshine and fresh air: A practical approach to combating waterborne disease. *Waterlines* 15:27–29.

Reed, R.H. (1997b). Solar inactivation of faecal bacteria in water: The critical role of oxygen. *Lett. Appl. Microbiol.* 24:276–280.

Reed, R.H., Mani, S.K., and Meyer, V. (2000). Solar photo-oxidative disinfection of drinking water: Preliminary field observations. *Lett. Appl. Microbiol.* 30:432–436.

Ribeiro, H. (2000). Diarrheal disease in a developing nation. *Am. J. Gastroenterol.* 95:S14–S15.

Rice, E.W., and Hoff, J.C. (1981, September). Inactivation of *Giardia lamblia* cysts by ultraviolet irradiation. *Appl. Environ Microbiol.* 42(3):546–547.

Saito, T., Iwase, T., Horie, J., and Morioka, T. (1992). Mode of photocatalytic bactericidal action of powdered semiconductor TiO_2 on mutans streptococci. *J. Photochem. Photobiol. B Biol.* 14:369–379.

Salih, F.M. (2002). Enhancement of solar inactivation of *Escherichia coli* by titanium dioxide photocatalytic oxidation. *J. Appl. Microbiol.* 92:920–926.

Schink, T., and Waite, T.D. (1980). Inactivation of *f2* virus with ferrate ŽVI. *Water Res.* 14:1705–1717.

Schuler, P.F., Ghosh, M.M., and Gopalan, P. (1991). Slow sand and diatomaceous earth filtration of cysts and other particulates. *Water Res.* 25(8):995–1005.

Shin, J.K., and Pyun, Y.R. (1997). Inactivation of *Lactobacillus plantarum* by pulsed-microwave irradiation. *J. Food Sci.* 62:163–166.

Slade, J.S., Harris, N.R., and Chisholm, R.G. (1986). Disinfection of chlorine-resistant enteroviruses in groundwater by ultraviolet radiation. *Water Sci. Technol.* 189(10):115–123.

Snider, K.E., Darby, J.L., and Tchobanoglous, G. (1991). *Evaluation of ultraviolet disinfection for wastewater reuse applications in California.* Davis, CA: Department of Civil Engineering, University of California–Davis.

Sobsey, M.D. (2002). *Managing water in the home: Accelerated health gains from improved water supply.* Geneva: World Health Organization. Available at http://www.who.intwater_sanitation_health/DocumentsWS0207/managingwater.htm (accessed September 26, 2003).

Sun, D.D., Tay, J.H., and Tan, K.M. (2003). Photocatalytic degradation of E. coliform in water. *Water Res.* 37:3452–3462.

Tchobanoglous, G.T. (1997). UV disinfection: An update. Presented at Sacramento Municipal Utilities District Electrotechnology Seminar Series, Sacramento, CA.

Trussell, R.R., and Chao J.L. (1977). Rational design of chlorine. *Contact Facilities* 49:659–667.

Tyrell, R.M. (1976). Synergistic lethal action of ultraviolet-violet radiations and mild heat in *Escherichia coli*. *Photochem. Photobiol.* 24:345–351.

USEPA. (1980). *Technologies for upgrading existing and designing new drinking water.*

USEPA. (1986). *Design manual: Municipal wastewater disinfection.* EPA/625/1-86/021. Office of Research and Development, Water Engineering Research Laboratory.

USEPA. (1996). *Ultraviolet light disinfection.*

U.S. NCI. (1976). Report on the Carcinogenesis Bioassay of Chloroform (CAS No. 67-66-3). TR-000. NTIS Rpt. No. PB264018. Bethesda, MD: National Cancer Institute.

Van der Kooij, D., Hijnen, W.A.M., and Kruithof, J.C. (1989). The effects of ozonation, biological filtration and distribution on the concentration of easily assimilable organic carbon (AOC) in drinking water. *Ozone: Sci. Eng. J. of the International Ozone Assoc.* 11(3):297–311.

Vicars, S. (1999). Factors affecting the survival of enteric bacteria in saline waters. Ph.D. thesis, Northumbria University, Newcastle upon Tyne.

Watanabe, K., Kakia, Y., Kashige, N., Miake, F., and Tsukiji, T. (2000). Effect of ionic strength on the inactivation of microorganism by microwave irradiation. *Lett. Appl. Microbiol.* 31:52–56.

Watts, R.J., Kong, S., Orr, M.P., Miller, G.C., and Henry, B.E. (1995). Photocatalytic inactivation of coliform bacteria and viruses in secondary wastewater effluent. *Water Res.* 29:95–100.

Wegelin, M., Canonica, S., Mechsner, K., Fleischmann, T., Pesaro, F., and Metzler, A. (1994). Solar water disinfection: Scope of the process and analysis of radiation experiments. *J. Water SRT-Aqua* 43:154–169.

Wegelin, M., and De Stoop, C. (1999). Potable water for all: Promotion of solar water disinfection. In *Proceedings of the Twenty-Fifth Water, Engineering and Development Centre Conference*, Addis Ababa, Ethiopia, pp. 310–312.

Wei, C., Lin, W., Zainal, Z., Williams, N.E., Zhu, K., Kruzic, A.P., Smith, R.L., and Rajeshwar, K. (1994). Bactericidal activity of TiO_2 photocatalyst in aqueous media: Toward a solar-assisted water disinfection system. *Env. Sci. Technol.* 28(5):934–938.

Welt, B.A., and Tong, C.H. (1993). Effect of microwave radiation on thiamine degradation kinetics. *J. Microwave Power Electromagnetic Energy* 28:187–195.

White, G.C. (1992). *Handbook of chlorination and alternative disinfectants.* New York: Van Nostrand Reinhold.

Wolfe, R.L. (1990). Ultraviolet disinfection of potable water. *Environ. Sci. Technol.* 24(6):768–773.

World Health Organisation International Agency for Research on Cancer. (1991). Chlorinated drinking water; chlorination by-products; some other halogenated compounds; cobalt and cobalt compounds. *IARC Monographs on the Evaluation of Carcinogenic Risks to Humans* 52:45.

Würtele, M.A., Kolbe, T., Lipsz, M., Külberg, A., Weyers, M., Kneissl, M., and Jekel, M. (2011). Application of GaN-based ultraviolet-C light emitting diodes—UVLEDs—for water disinfection. *Water Res.* 45(3):1481–1489.

Xu, X., Zhou, H., He, P., and Wang, D. (2005). Catalytic dechlorination kinetics of p-dichlorobenzene over Pd/Fecatalysts. *Chemosphere* 58:1135–1140. Available at http://dx.doi.org/10.1016/j.chemosphere.2004.07.010 (accessed August 15, 2012).

Yahya, M.T., Landeen, L.K., Forsthoefel, N.R., et al. (1989). Evaluation of potassium permanganate for the inactivation of MS-2 in water systems. *J. Environ. Sci. Health* A34(8):979–989.

Yip, R.W., and Konasewich, D.E. (1972). Ultraviolet sterilization of water—Its potential and limitations. *Water Pollut. Control (Canada).* 14:14–18.

4

Hybrid Techniques

4.1 Introduction

The combination of drugs in order to obtain an effect greater than that of any one compound taken alone has been practiced for generations (Dahi, 1976), and this has also been applied to water treatment. Multiple disinfectants, the sequential or simultaneous use of two or more disinfectants, have been used with increasing success in recent years. This trend is attributed to the fact that the water industry is pressed to look for more effective disinfectants due to continuing pressure from the regulatory bodies as well as the consumers. More effective disinfection can be achieved by higher doses of disinfectants. However, this also leads to higher amount of disinfection by-products (DBPs), which itself can be a major health concern. Recent research has shown that the application of sequential disinfectants is more effective than the added effect of the individual disinfectants. Therefore, hybrid (combo) or synergistic techniques for water disinfection appear to be an apt solution for the current challenges faced in the field of water disinfection.

4.1.1 What Is the Basis of a Hybrid Method?

The majority of chemical as well as a few physical water disinfection methods, such as UV and cavitation, are known to generate hydroxyl radicals that are the sole disinfecting species. Therefore, an increase in their generation would naturally augment the disinfection process. Several oxidation techniques are based on this principle, and these are known as the advanced oxidation processes (AOPs). Hybrid techniques employ a combination of various oxidation techniques, which results in an accelerated production of hydroxyl radicals to affect water disinfection. The hydroxyl radical is a powerful oxidising radical, second only to fluorine (Paul and Canter, 1990), and is therefore appropriate to use in water treatment, as it does not produce any unwanted residuals. Advanced oxidation systems generally combine ozone, hydrogen peroxide, and ultraviolet radiation, for example, O_3 and H_2O_2, O_3 and UV, and H_2O_2 and UV.

4.1.2 Inactivation Mechanism

Bernbaum (1981, 1985) developed a testing method for determining the kind of interaction that can be expected when agents are combined to produce a given observation. Synergism can be tested using the mathematical model developed by Bernbaum and modified for disinfection kinetics by Kouame and Haas (1991). The principle is that if the agents in a given combination do not interact in producing the effect observed, then regardless of the effect relations, the following equation is satisfied:

$$\frac{n}{n-1} \sum \frac{x_i}{y_i} = 1$$

< 1	Synergetic
> 1	Antagonistic
= 1	Additive

where x_i and y_i are the concentration of the individual agent present in the combination and the concentration of the agents that individually would produce the same magnitude of effect as that of the combination, respectively, i is the individual agent, and n is the total number of agents. The sum calculated from this equation for a set of data is interpreted as a case of synergistic action if the sum is less than 1, antagonistic interaction if the sum is greater than 1, and additively, that is, no interaction if the sum is equal to 1.

4.1.3 Factors Affecting Hybrid Disinfection

Similar to most chemical disinfectants, pH and temperature affect the amount of synergistic inactivation achieved by sequential applications of disinfectants (Finch, 1997). The level of inactivation due to the sequential application of chemical disinfectants is believed to be pH dependent (Finch, 1997). The impact of pH on the log inactivation of *Cryptosporidium parvum* attributed to synergistic effects for three sequential combinations of ozone–chlorine dioxide, chlorine dioxide–free chlorine, and chlorine dioxide–chloramine, respectively. The amount of log inactivation due to synergistic effects is lower at high pH (e.g., pH = 11). These results show that neutral pH is more effective than low pH except for the ozone–chlorine dioxide combination. The combined effect of low temperature and high pH is believed to significantly reduce the amount of *Cryptosporidium* inactivation attributed to synergism (Finch, 1997). One possible explanation for this reduction is that the oocysts contract under these conditions and become harder to penetrate. However, significant reduction in *Cryptosporidium* oocyst inactivation is true under reduced water temperature and high pH whether interactive disinfection is practiced or not. Therefore, reduced inactivation may not be necessarily due to synergism between combined disinfectants. *Cryptosporidium* oocysts are more susceptible to inactivation by combinations of disinfectants than by individual disinfectants. However, no synergism was observed with bacterial spores, specifically *Bacillus cereus* spores (Finch, 1997). These results suggest that encysted parasites might show more susceptibility to synergistic effects than bacterial spores. Masking effects caused by turbidity for interactive disinfectants are expected to be similar to those of the individual disinfectants.

4.1.4 Primary and Secondary Disinfectants

It is not only essential that water is effectively disinfected, but it should also remain so until it reaches the consumer. The latter is generally taken care of by residual disinfectants, which function by preventing microbial regrowth in the distribution networks delivering the required quality of water to the end user. Here, disinfection for two separate purposes can be recognised: primary and secondary disinfection. Primary disinfection refers to the inactivation of microorganisms to meet the regulatory bacteriological requirements, which are met by achieving certain contact time (CT) requirements to ensure a target log inactivation of the microorganisms as set forth by the regulating water authorities. Secondary disinfection generally refers to the application of a disinfectant to meet regulatory requirements for the distribution system's bacteriological quality. To be an effective primary disinfectant, it should effectively inactivate the target microorganism. As discussed in earlier chapters, some disinfectants, such as ozone, UV, and peroxone, while being effective disinfectants, do not leave a long-lasting residual. Therefore, secondary disinfection becomes imperative, and its selection also becomes limited to those disinfectants that remain stable in the distribution system.

4.2 Hybrid Disinfection Techniques

4.2.1 Advanced Oxidation Processes

These are hybrid processes aimed to augment water disinfection by generating highly reactive hydroxyl radicals, which inactivate microorganisms, rendering the water potable. Advanced oxidation processes often make use of UV irradiation since the effectiveness of oxidising agents like hydrogen peroxide, chlorine, and ozone is enhanced in the presence of UV irradiation. UV light, when combined with an oxidising agent, provides extra energy to boost the free radical formation from either ozone or hydrogen peroxide and oxidises organic pollutants to simpler, less refractory organic compounds or to carbon dioxide, water, and mineral acids (Paul and Canter, 1990). UV radiation also keeps the organic species in the excited state and makes them more vulnerable to the attack by the free radicals. Several biorefractory compounds can be effectively oxidised by photochemical oxidation, as their biodegradability is often enhanced after UV treatment (Paul and Canter, 1990). It is also interesting to note that most microbes are sensitive to lower concentrations of oxidising agents when exposed to ultrasound and the combined action of UV radiation, and the high-frequency ultrasound increases the rate of bacterial inactivation (Sierra and Boucher, 1971).

The peroxone process is one of the several advanced oxidation processes that have been tried for water disinfection. The peroxone process typically employs either UV or hydrogen peroxide to accelerate the decomposition of ozone, thereby generating highly reactive hydroxyl radicals. Oxidation in the peroxone occurs due to two reactions: direct oxidation of compounds by dissolved ozone and oxidation of compounds by hydroxyl radicals produced by the decomposition of ozone (Hoigné and Bader, 1978). It is important to understand that in the case of direct ozonation, only the first mode of disinfection usually operates. According to Hoigné and Bader, there are several features of peroxone that distinguishes it from only ozonation, and these are summarised in Table 4.1.

Because the radical oxidation is much more effective than the direct oxidation with ozone, it has been used extensively to treat difficult-to-oxidise organics

TABLE 4.1

Major Differences between the Peroxone and Ozonation Techniques for Water Disinfection

Serial No.	Ozonation	Peroxone
1	The ozone decomposition is generally slower than the peroxone process. Normally ozone decomposes, forming a hydroxyl radical as the intermediate.	The ozone decomposition rates are enhanced, leading to a higher concentration of hydroxyl radicals than that obtained with only ozone.
2	The ozone residual concentration is generally higher than the peroxone process.	The residual concentration of hydroxyl radicals is relatively small.
3	The process relies on direct oxidation by dissolved oxygen, which is less reactive than hydroxyl radicals.	Here, the oxidation is by hydroxyl radicals generated, which is much more reactive than only ozone.
4	It can be considered a predisinfectant since an ozone residual can be determined.	Peroxone may not be an appropriate option as a predisinfectant since the residual measurement is very difficult.
5	The oxidation potential of ozone is less than that of the hydroxyl radicals.	Hydroxyl radicals have a higher oxidation potential than ozone and are much more reactive.

such as taste and odour compounds and chlorinated organics such as geosmin, 2-methylisoborneol (MIB), phenolic compounds, trichloroethylene (TCE), and perchloroethylene (PCE), to name a few (Pereira and Phelps, 1996; Ferguson et al., 1990). The optimum peroxide:ozone dose ratio to maximise hydroxyl radicals' reaction rate can be determined for a specific oxidation application. For instance, the optimum peroxide:ozone dose ratio for TCE and PCE oxidation in groundwater was determined to be 0.5 by weight (Glaze and Kang, 1988). Tests showed that TCE required lower ozone dosages for the same percentage removal than PCE. pH and bicarbonate alkalinity also play a major role in peroxone effectiveness (Glaze and Kang, 1988). This is primarily related to bicarbonate and carbonate competition for the hydroxyl radical at high alkalinity, and carbonate competition for hydroxyl radical at high pH levels. However, turbidity alone does not appear to play a role in peroxone effectiveness, nor does peroxone appear to remove turbidity (Tobiason et al., 1992). A study conducted at the Metropolitan Water District of Southern California (MWDSC) indicated that the performance of peroxone is greatly dependent upon the peroxide:ozone ratio (Wolfe et al., 1989). Results from previous studies at MWDSC suggested that the optimal ratio for disinfection was less than or equal to 0.3.

The peroxone process requires an ozone generation system and a hydrogen peroxide feed system. The process involves two essential steps: ozone dissolution and hydrogen peroxide addition. Hydrogen peroxide can be added after ozone, before ozone, or simultaneously. The most efficient operation is to add ozone first to obtain CT disinfection credit, followed by peroxide for hydroxyl radical oxidation. Addition of hydroxyl free radicals initially should be minimised since the hydrogen peroxide competes with ozone-reactive molecules (i.e., initial demand) for the ozone present. In the second stage, organic matter is already oxidised; therefore, adding hydrogen peroxide during the second stage makes it possible to raise the overall oxidation efficiency (Duguet et al., 1987).

4.2.1.1 How Does Peroxone Inactivate Pathogens?

Both peroxone and other advanced oxidation processes have been proven to be equal or more effective than ozone for pathogen inactivation. Disinfection credits are typically described in terms of CT requirements. Because peroxone leaves no measurable, sustainable residual, calculation of an equivalent CT for disinfection credit is not possible unless there is a measurable ozone residual. The primary cause for the pathogen inactivation is attributed to ozone, specifically the mechanisms associated with the oxidation of pathogens by direct ozone reaction and by the hydroxyl radicals. Studies using a combination of ozone–hydrogen peroxide have shown that the disinfection of *E. coli* is less effective, as the peroxide-to-ozone ratio increases to above (approximately) 0.2 mg/mg (Wolfe et al., 1989). The decrease in the disinfection was believed to be caused by lower ozone residuals associated with higher peroxide-to-ozone ratios, which indicates that the direct ozone reaction is an important mechanism for pathogen inactivation. A study conducted by Ferguson et al. (1990) compared the pathogen inactivation capability of peroxone and ozone using MS-2 and f2 coliphages as well as *E. coli* and heterotrophic plate count (HPC) bacteria. The f2 and MS-2 coliphages were comparable in their resistance to ozone and peroxone.

4.2.1.2 Disinfection By-Products of the Peroxone Process

The principal by-products associated with peroxone are similar to those formed by the ozonation. The MWDSC study found that the use of peroxone/chlorine resulted in trihalomethane (THM) concentrations 10 to 38% greater than the use of ozone/chorine. However, the THM concentrations of waters disinfected with peroxone/chloramines and ozone/chloramines were similar (Ferguson et al., 1990). The use of peroxone as a primary disinfectant and chloramines as a secondary disinfectant can successfully control halogenated DBP formation if bromide ion is not present and adequate CT credit can be established. It has been shown that peroxone produces more bromate ion than ozone when similar ozone residuals (CT credits) are achieved (Krasner et al., 1993). A study by MWDSC evaluated the effectiveness of peroxone to control taste and odour, DBPs, and microorganisms (Ferguson et al., 1990). The study found that the two sources of waters disinfected with peroxone, with free chlorine as the secondary disinfectant, resulted in THM concentrations ranging from 67 to 160 mg/l. Conversely, using chloramines as a secondary disinfectant resulted in THM concentrations consistently below 3.5 mg/l (Ferguson et al., 1990).

4.2.2 Free Chlorine and Monochloramine

Kouame and Haas (1991) studied the inactivation of *E. coli* in a continuous flow system to a continuous stirred tank reactor (CSTR) in which free chlorine and monochloramine simultaneously exist. They found that the interaction between free chlorine and monochloramine while reacting with *E. coli* was synergistic and, when free chlorine and monochloramine exist in a continuous flow system, high inactivation of bacteria occurred. They explain the mechanism of this synergism between free chlorine and monochloramine as follows: free chlorine is capable of causing sublethal injury to bacteria (Camper and McFeters, 1979), and these injured bacteria are more sensitive to monochloramine. This work brings out the efficacy of synergism in the inactivation of microorganisms. Another hypothesis is that the first oxidant (i.e., chlorine, chlorine dioxide, or ozone) conditions the outer membrane of *Cryptosporidium* oocysts so that the secondary oxidant (i.e., chlorine, chlorine dioxide, and monochloramine) can penetrate the oocyst more easily (Liyanage, Finch, and Belosevic, 1997). For example, preliminary work on the disinfection of *Cryptosporidium parvum* using free chlorine, followed by monochloramine, suggested that there may be a synergism involving two chlorine species. Sequential treatment of these chlorine species was found to provide greater inactivation than expected from the additive effects of the two disinfectants used alone (Gyurek et al., 1996). Studies have utilised a straightforward method to determine if synergism has occurred based on measured inactivation (Finch, 1997; Gyurek et al., 1996; Liyanage, Finch, and Belosevic, 1997). According to this approach, synergism is demonstrated if the sequential application of the disinfectants provides more inactivation than is expected from the additive effects of the individual, separate disinfectants. In addition, the magnitude of the synergistic effects is equal to the difference in the level of inactivation achieved from multiple disinfectants and the additive inactivation achieved from the single disinfectants. Work on the mechanism of inactivation of bacteria by this synergism has led to some detailed study, and it appears that the disinfection is due to the damage to the nucleic acids of the bacteria. Interestingly, only small differences in the rate of inactivation between the high nucleic acid content bacteria and low nucleic acid content bacteria were observed for the combination of chlorine and monochlorine, in contrast to the major differences seen for other synergisms, such as ozone and chlorine dioxide and ozone and chlorine (Ramseier et al., 2011). Like any other chemical disinfection, the combination of free chlorine and monochloramine also results in the generation of DBPs. Typically, trihalomethanes (THMs), haloacetic acids (HAAs), halonitromethanes (HNMs), haloacetonitriles (HANs), haloaldehydes (HAs), haloketones (HKs), and iodo-THMs (i-THMs) are some of the commonly reported DBPs. It was also found that compared to chlorine, treatment of water with monochloramination generally resulted in lower concentrations of DBPs with the exception of 1,1-dichloropropanone (Bougeard et al., 2010).

4.2.3 Metal Ions and Chemical Disinfection

Copper and silver ions have been used for many years for the disinfection of water. Copper and silver are known to affect a number of microorganisms, including bacteria, viruses, and algae (McFeters and Singh, 1991). They are believed to interfere with the enzymes involved in cellular respiration (Domek et al., 1984) and to bind at specific sites to deoxyribonucleic acid (DNA). Inactivation by combined copper and silver has been shown to be relatively slow when compared with that of free chlorine. However, when these metals were added to water with low levels of free chlorine, inactivation rates of bacterial indicator organisms were shown to be greater than those at comparable levels of free chlorine alone (Yayha et al., 1989). Landeen et al. (1989) evaluated water distribution systems utilising electrolytically generated copper and silver ions and low levels of free chlorine at room and elevated temperatures in filtered well waters for their efficacy in inactivating *Legionella pneumophila*. They report increased inactivation rates when 400 and 40 µg of copper and silver per litre were added at 0.4 mg of free chlorine per litre.

4.2.4 Ozonation and Ultraviolet Radiation

The potential of both ozone and ultraviolet radiation for water disinfection has already been discussed in detail in the preceding chapters. In this section, the synergism of these two processes is elaborated. It is well established that high-energy UV radiation can break covalent bonds in organic compounds and disrupt molecules, leading to the production of free radicals and ions. Combined use of ozone and UV results in the decomposition of ozone to the ozonide radical, O_3^-, which is a precursor to the hydroxide radical. This can then result in the formation of hydrogen peroxide if the radicals are sufficiently close in proximity. Typically, lamps emitting at about 254 nm are used for photocatalytic oxidation (Clayton, 1992). The general UV/ozonation reactions in the aqueous phase are given in the following equation (Paul and Canter, 1990):

$$O_3 + H_2O + hv \rightarrow 2(OH) + O_2$$

Kusakabe et al. (1990) noted that carcinogenic and mutagenic activity of chlorinated organic compounds has become of major concern with regard to human health. Ozone is a strong oxidant, but direct attack by molecular ozone is not suitable for the complete oxidation of the dissolved organic compounds because of the formation of the refractory compounds, which accumulate in the solution. The reaction involving the OH radical formed by the photocatalytic decomposition of ozone is fast and nonselective (Ikemizu et al., 1987, in Kusakabe et al., 1990). The combination of UV radiation and ozonation for the treatment of humic acids was studied by Kusakabe et al. (1990), where the destruction rate of humic acids was determined by separating the contribution of UV radiation from that of the ozonation. It was found that UV radiation enhanced the total organic carbon destruction rate. Takahashi (1990) studied the efficacy of ozonation/UV treatment for low molecular weight organic compounds. The oxidation of phenol was promoted by the simultaneous use of ozone and UV. This method has been recommended as a suitable advanced treatment method. It has been well established by a number of studies that the combination of ozonation at reduced doses and UV treatment leads to an improved water quality with regard to disinfection, oxidation of micropollutants, and minimisation of bromated DBPs (Meunier et al., 2006).

4.2.5 Ultrasound and Ozone

Degradation of aromatic compound in water using a combination of sonolysis and ozonolysis has been investigated by Weavers and Hoffmann (1998). They found an enhancement in the loss of total organic carbon (TOC) by sonolytic ozonation. Burleson et al. (1975) investigated the combination of ozone and ultrasound for the inactivation of microorganisms. It was found that sonication alone did not inactivate the bacteria, but the combined effect of ozonation and sonication enhanced the efficacy of the treatment, resulting in a synergistic effect. It was found that cavitation was produced by sonication, which enhanced the inactivation by reducing the high surface tension caused by organic material. Cavitation also breaks bubbles, resulting in small bubbles and an increased interfacial area exposed to ozone, increasing its dissolution rates.

Dahi (1976) studied the disinfection of water by means of ultrasound and ozone. Ultrasound at a frequency of 20 kHz was used in combination with ozone diluted in air at a dose of 60 ml/min to disinfect effluent from a biological sewage plant. The efficacy of ozone was enhanced by simultaneous treatment with ultrasound, and then enhanced further by sonication prior to ozonation. A number of mechanisms have been reported:

1. The most common explanation for the influence of ultrasonics is the theory of the desegregation of flocs of microorganisms. This concept has been accepted by various authors (Katzenelson et al., 1974, in Dahi, 1976; Burleson et al., 1975).

2. Another hypothesis is that of Kryszczuk (1962, in Dahi, 1976), who reports a transient rupture of chemical bonds between molecular components of cellular membranes, which results in an increase in the permeability of disinfecting substances in general.

3. Boucher et al. (1967) assume ultrasonic acceleration of diffusion, allowing more rapid penetration of the toxic gas molecule into the microorganism.

4. Dahi (1976) states that the disinfectant and oxidant of ozonation are the free radicals, which are produced when ozone decomposes. Ultrasonic treatment increases the ozone decomposition and the activity of free radicals in water. Ultrasonic treatment also increases the aeration parameter. When the activity of free radicals is attained, a very rapid inactivation of bacteria is observed (Dahi, 1976).

4.2.6 Ultrasound and Chemical Disinfection

The combination of ultrasound and chemical disinfection was studied by Sierra and Boucher (1971). Inactivation of high-density bacterial spore suspensions was achieved by the treatment with low concentrations of aqueous acid glutaraldehyde solutions at temperatures above or about 54°C with ultrasound. Both low (20 kHz) and high (250 kHz) frequencies were found to be synergistic with glutaraldehyde by a reduction in the time required by alkaline or acid aqueous glutaraldehyde solutions to inactivate spores of *Bacillus subtilis* at all the temperatures studied. They found that ultrasonic energy did not sensitise the spores to glutaraldehyde, because sonicated spores were as resistant to glutaraldehyde as intact spores. They suggest that synergistic phenomena between ultrasound and glutaraldehyde occurred.

High-frequency ultrasound alone resulted in the spores gradually losing their viability, leaving empty spores, spore coat fragments, and other insoluble components. Two important factors in the inactivation of bacterial spores by a chemical are:

1. The chemical must penetrate the spore.
2. The chemical must react with the spore site(s) of inactivation.

Their studies revealed that the spore is permeable to glutaraldehyde regardless of pH and temperature, and that the chemical penetrates the spore almost at once. They state that the potential of the glutaraldehyde inactivation of bacterial spores with ultrasonic energy is the result of an increase in the rate of chemical reaction between the chemical and the spore site.

When hydrogen peroxide is added to water before subjecting it to ultrasonication in the bath or the horn, the overall disinfection rate obtained is higher than that obtained when only the ultrasonic horn or the ultrasonic bath is used (Jyoti and Pandit, 2003). This can be explained as follows:

1. One hypothesis is that ultrasound ruptures the chemical bonds between molecular components in the cell membranes of the microorganisms, which leads to an increase in the permeability to chemical substances like hydrogen peroxide (Dahi, 1976).
2. Finally, the increase in the efficacy of this hybrid process may also be due to an increase in the hydrogen peroxide decomposition and the activities of free radicals in water by the ultrasonic treatment (Dahi, 1976), though no specific evidence of increased H_2O_2 decomposition rate has been observed.

It has also been observed that the overall disinfection rates obtained for the combination of ultrasonication and hydrogen peroxide were more than additive in the case of the HPC bacteria, as well as the indicator microorganisms studied (Jyoti and Pandit, 2003). Continuing research in this area led to interesting findings and relations between different sonication parameters and the effect of the chemical disinfectant on inactivation of water pathogens. It is remarkable to note a frequency effect when ultrasound was used as both pretreatment and simultaneous treatment with sodium hypochlorite to inactivate *E. coli*. At a lower frequency of 20 kHz, the biocidal activity was greatest when ultrasound was applied simultaneously. On the other hand, at higher ultrasonic frequency of 850 kHz the biocidal effects were greatest when sonication was used as a pretreatment immediately followed by hypochlorite treatment. Although the microbicidal rates were very similar for both cases, pretreatment at higher frequency appears to be promising, as it involves less acoustic energy, and hence energetically is more efficient (Duckhouse et al., 2004). Studies were extended to investigate the effect of a photocatalyst such as TiO_2 on the efficiency of ultrasound. Interestingly, TiO_2 was found to greatly improve the disinfection process. A 98% reduction in the concentrations of viable cells of *E. coli* was observed in the presence of TiO_2 during a 30 min period of treatment with combined ultrasound and titanium dioxide, while only a 13% reduction was observed when only ultrasonic irradiation was used. The study also clearly confirmed the role of hydroxyl radicals in the inactivation mechanism of the synergistic combination (Dadjour et al., 2005). Very recently, wastewater treatment using ultrasound as a pretreatment before the application of chlorine dioxide revealed that the combined use was synergistic, involving high removal (3.2 to 3.5 log reduction) of *E. coli* and total coliform from the raw wastewater. The individual treatments only provided 1.4 to 1.9 log reductions. The enhanced effect obtained with the hybrid process was attributed to the breaking up of the clustered particles in the raw wastewater by the sound waves, thereby exposing the microbes to chlorine dioxide (Ayyildiz et al., 2011). Similar trends can be expected in the case of drinking water also.

4.2.7 Hydrogen Peroxide and Ultraviolet Radiation

The rate of oxidation can be accelerated by the addition of hydrogen peroxide to water and then subjecting it to UV irradiation since it is known that cleavage of the O-O bond in hydrogen peroxide gives rise to hydroxyl radicals (Clarke and Knowles, 1982). Baxendale (1957) in Clarke and Knowles (1982) reports that photodecomposition of hydrogen peroxide proceeds via a chain reaction initiated by the hydroxyl radical after absorption of light at wavelengths up to 365 nm (UV range):

$$H_2O_2 + h\nu \rightarrow 2OH\cdot$$

$$OH\cdot + H_2O_2 \rightarrow H_2O + HO_2\cdot$$

$$HO_2\cdot + H_2O_2 \rightarrow H_2O + O_2 + OH\cdot$$

$$2HO_2\cdot \rightarrow H_2O_2 + O_2$$

The hydroxyl radicals preferentially attack the organic compounds by abstracting a hydrogen atom from the molecule, resulting in the formation of organic radicals that themselves can initiate several chain reactions (Walling, 1975, in Clarke and Knowles, 1982). Hydrogen peroxide produces the highest concentration of radicals per mole of oxidant, therefore making it an obvious choice in the design of a cost-effective oxidising system. The production of hydroxyl radicals from hydrogen peroxide requires a dissociation energy of 48.5 kcal/mol, which requires shortwave UVC wavelengths (from

200 to 280 nm) (Clarke and Knowles, 1982). The molar extinction coefficient of hydrogen peroxide at 254 nm is only 19.6/M/cm. This is low compared to the molar extinction coefficient of ozone, which is 3300/M/cm. The implication of this is that a high concentration of hydrogen peroxide is required to generate sufficient levels of OH radicals.

It has been noted that employing only hydrogen peroxide for water disinfection implies a higher concentration of the same for the destruction of pathogenic organisms, which can lead to phytotoxicity and also result in higher concentrations of disinfection by-products. Therefore, synergistic use of hydrogen peroxide along with UV radiation is known to enhance the disinfection efficiency and also reduce the amount of chemical required for inactivation. Solar UV rays can also bring about this effect. This was observed in the case of inactivation of *Fusarium solani*. Experiments conducted in laboratory and pilot plant solar reactors employed both distilled and real well water under natural sunlight. This was the first study that showed the promising disinfection ability of combined hydrogen peroxide along with solar UV rays for fungal disinfection (Sichel et al., 2007). It is also interesting to note that apart from disinfection, UV and hydrogen peroxide can be effectively used to control odourants and brominated disinfection by-products (Jo et al., 2011).

4.2.8 Hydrogen Peroxide, Ultraviolet Radiation, and Hydrodynamic Cavitation

The Cav-ox system employs the combination of hydrogen peroxide, hydrodynamic cavitation, and UV irradiation. Reports on the Cav-ox system indicate a 100% reduction of bromoform, an 89.3% reduction of phenol, and a 100% reduction of O-chlorophenol from industrial wastewater. These results were obtained with an addition of hydrogen peroxide ranging from 50 to 100 mg/l. A 99.99% reduction of bacteria in drinking water was achieved with the addition of 150 mg/l of hydrogen peroxide (Neytzell-de Wilde and Chetty, 1990). It is claimed that free radicals, mainly OH radicals, are generated by cavitation, which results in the thermal dissociation of water, and by the splitting of hydrogen peroxide by UV irradiation. The free radicals are said to split organic molecules in a chain reaction. The individual effects of hydrogen peroxide, UV irradiation, and hydrodynamic cavitation are also the contributing factors.

4.2.9 Hydrogen Peroxide, Ultraviolet Radiation, and Iron Catalyst (Photo-Fenton Process)

The photo-Fenton process utilises the combination of $Fe(II)/H_2O_2/UVA$ or $Fe(III)$ oxalate/H_2O_2/UVA. A combination of H_2O_2 and UVA radiation with $Fe(II)$ or $Fe(III)$ oxalate produces more hydroxyl radicals in comparison with the systems $Fe(II)/H_2O_2$ or H_2O_2/UVC, thus promoting the rate of degradation of organic pollutants. The proposed reaction pathways for the photo-Fenton process start with the primary photoreduction of dissolved $Fe(III)$ complexes to $Fe(II)$ ions followed by the Fenton reaction and the oxidation of organic compounds. The hydroxyl radicals formed in the photo-Fenton process are highly reactive oxidants and initiate the oxidative destruction of organic substances (RH) in water, which may lead to a total mineralisation of the organic pollutants (Kim and Vogelpohl, 1998). Kim and Vogelpohl (1998) utilised the photo-Fenton process with success to degrade biorefractory organic pollutants in landfill leachate. They found that the rate of degradation of organic pollutants depends on the concentration of hydrogen peroxide and the iron catalyst, the pH value, and the concentration of the dissolved oxygen. They also compared the photo-Fenton process with the $H_2O_2/Fe(II)$ and H_2O_2/UVC processes and found that the photo-Fenton process gives a higher chemical oxygen demand (COD) degradation and reduced energy consumption of at least 30% than the H_2O_2/UVC process. They found that by using photogenerated $Fe(II)$, the amount of iron catalyst required and the volume of sludge produced is substantially reduced. Although there are no reports on the use of the photo-Fenton process for disinfection of water, this process has a lot of potential to be used for the inactivation of microbes in water because metals like iron are known to be detrimental to the microbial cell, and the process stated for the degradation of organic pollutants can apply for the microbes as well. However, it may also lead to an increase in the total dissolved solids (TDS) of the treated water owing to the presence of iron salts.

4.2.10 Photocatalysis

Photocatalytic disinfection of municipal wastewater was studied by Dillert et al. (1998). An advanced sewage treatment was performed in the scope of (1) the chemical conversion of toxic organic pollutants to less harmful, biodegradable compounds, and (2) the reduction/inactivation of bacteria and other pathogenic microorganisms. Thus, municipal wastewater detoxification and disinfection was accomplished coincidently by photoinitiated advanced oxidation process. Photocatalysis employs low-cost titanium dioxide as the heterogeneous photocatalyst, water-dissolved molecular oxygen as the oxidant, and photons of the ultraviolet part of the solar spectrum as the source of energy for the oxidation of organic compounds and for the disinfection (Ireland et al., 1993). The underlying reactions are initiated as the semiconductor photocatalyst absorbs ultraviolet light; that is, solar photons of wavelengths below 380 nm, which will generate electron-hole pairs. Once these charged carriers have migrated to the surface of the photocatalyst, they are able to undergo redox reactions with those molecules and ions that are adsorbed on the solid surface. Generally, the photocatalytic reduction of the pollutants and the bacteria are explained by the oxidising action of the photogenerated electron-holes and surface-bound hydroxyl radicals (Dillert et al., 1998). Dillert et al. (1998) report that the disinfection is a direct consequence of both the direct action of light on the microorganism and the photocatalytic action of the excited photocatalytic particles. They also state that the organic pollutants present in the municipal wastewater and their transformation products, formed by photolysis or photocatalysis, can also act as germicides, in addition to the primary mechanisms discussed earlier. In another study, the potential application of TiO_2 photocatalysis as the primary disinfection system of drinking water was studied in terms of inactivation of coliform bacteria; 91 to 99% inactivation was reported after 60 min of irradiation, which also depended on the loading of the catalyst and the initial number density of coliforms. Interestingly, it was noted that bacterial damage was not irreversible and regrowth was possible under optimal conditions if final disinfection was not employed (Rizzo, 2009). Yet another study compares the efficacy of TiO_2/UV and $Ag–TiO_2/UV$ systems for the inactivation of *Giardia intestinalis* and *Acanthamoeba castellani* cysts. Silver-loaded TiO_2 was found to be much more effective than the TiO_2/UV system, and this was due to the better catalysis as well as antimicrobial activity of silver. The cell wall of the cysts was found to be irreversibly damaged and did not regrow (Sökmen et al., 2008).

4.2.11 Mechanical and Chemical Disinfection

Parker and Smith (1993) studied the effect of shaking *C. parvum* oocysts with sand and the effect of chlorination on sand-damaged *C. parvum* oocysts. They found that shaking oocysts with sand for 5 min, 90 min, and 2 h increased the number of nonviable oocysts to 50, 99.7, and 100%, respectively. Increasing the duration of shaking caused nonviable oocyts to lose their contents and fragment. They found that agitation with sand for 5 min and subsequent chlorination for 5 min increased the number of nonviable oocysts to 68.02%. Shaking oocysts with sand causes collision between oocysts and sand grains, which is detrimental to the oocysts' viability. In addition, the exposure of the oocysts to sand may be sufficient to render them susceptible to low concentrations of free chlorine, which have been shown to be ineffective on oocysts isolated from human faeces (Smith et al., 1989). The potential of mechanically mixed chemical disinfectants for bacterial inactivation continues to enthrall researchers. In an interesting investigation, the mechanically mixed oxidants containing Cl_2/O_3, Cl_2/ClO_2, and Cl_2/ClO_2^- were compared with Cl_2 alone for inactivation of *Bacillus subtilis* spores (indicator organism). Enhanced disinfection efficiency was seen in the mechanically mixed hybrid chemicals. This was due to the synergistic effect of the mixed oxidant itself and the effect of intermediates such as ClO_2^-/ClO_2, which were generated from the reaction between an excess of Cl_2 and a small amount of O_3/ClO_2^-. Mechanically mixed oxidants incorporating excess chlorine were proposed as a new and moderately efficient method of disinfection (Son, Lim, and Khim, 2009).

4.2.12 Heat and Oxidants

The use of an oxidant at an elevated temperature may be an effective control strategy. This strategy was used by Harrington et al. (1997) to control zebra mussels in water. Zebra mussels (*Dreissena polymorhpha*) are fresh water molluscs usually about the size of a fingernail but capable of growing to a length of about 5 cm. The adult mussels are generally about 0.68 to 3.8 cm in length. They are shelled and their shells have varied shape and are much larger than microorganisms. In engineered structures such as power plants and water intakes, zebra mussel colonisation can result in losses in hydraulic capacity, clogging of strainers and condenser tubes, obstruction of valves, and interference with service water and fire protection systems (Benschoten et al., 1993). Harrington et al. (1997) state that only thermal control has several disadvantages. This method is limited only to industries where excess heat is available; the water temperatures in some portions of a treatment plant (e.g., service water systems) cannot always be raised to temperatures high enough to kill mussels. Moreover, complications due to excess heating or abrupt thermal expansion of equipment are possible. Additionally, the high temperatures may kill fish or other aquatic organisms. Harrington et al. (1997) used chlorine or ozone to control zebra mussels at temperatures from 30 to 36°C. The study showed that the addition of chlorine or ozone was more effective than heat alone at test temperatures above 30°C. Compared to the heat alone, the combined use of heat and oxidants decreased the time to 95% mortality by more than 95% at 30°C. Above 30°C the benefits of the combined treatment strategy decreased with increasing test temperature. It has been reported that the addition of chlorine or ozone at elevated temperatures can reduce mortality times by as much as three orders of magnitude compared to oxidant addition at ambient temperatures.

4.3 DBP Formation with Various Primary and Secondary Disinfectant Combinations

The concentrations and types of DBPs formed depend on, among other things, the combination of disinfectants used to achieve primary and secondary disinfection and the water quality. For example, under certain water quality conditions, ozone/chloramine disinfection is known to produce lower THM concentrations than chlorine/chloramine disinfection. However, the ozone/chloramines alternative can increase the formation of other DBPs, such as aldehydes and biological organic matter (BOM). Only one of the disinfectants is applicable to all the situations. Raw water chlorination, applied prior to natural organic matter (NOM) removal processes, combined with chlorination for residual disinfection, produces the greatest concentrations of halogenated DBPs. Studies indicate that preoxidation of raw water with ozone or chlorine dioxide can reduce the formation of halogenated DBPs because it shifts the point of chlorine application from raw water to settled or filtered water, which has lower DBP precursor concentrations (MWDSC and JMM, 1989). The use of ozone can reduce the formation of halogenated by-products in waters containing low concentrations of bromide. However, ozone increases BOM and may encourage bacterial growth in the distribution system. Removal of activated organic carbon (AOC) with biological filtration (e.g., biological activated carbon) reduces the potential for bacterial growth in the distribution system. The use of chloramines as a secondary disinfectant, instead of chlorine, shortens the required chlorine contact time and thus reduces the formation of chlorinated by-products. However, chloramine produces by-products of its own (cyanogen chloride and cyanogen bromide). In addition, a short chlorine contact time prior to ammonia addition will help inactivate heterotrophic plate count bacteria that are found in the effluent of a biologically active filter. Bench or pilot studies will be required to evaluate the trade-offs in DBP formation for various disinfectant combinations for a specific application. The application of ozone should be carefully considered because it produces aldehydes, aldoketoacids, and carboxylic acids. However, these can be removed in a biologically active filter. In bromide-containing waters, ozonation can increase the formation of brominated organic DBPs and form bromate. In pilot plant studies for water containing low concentrations of bromide, Lykins et al. (1991) determined that ozonation followed by chloramination produced the lowest levels of halogenated disinfection by-products. However, this is not applicable to source waters containing significant bromide concentrations due to the potential for bromate formation and brominated THMs and HAAs. The addition of chlorine dioxide will produce chlorite and chlorate and may form some oxygenated DBPs (e.g., maleic acids).

4.4 Conclusion

Hybrid water disinfection techniques are definitely here to stay—for as long as life exists. This emphatic statement is backed by the discussions in the preceding sections and chapters, which have constantly pointed out that use of a hybrid technique is always favourable compared to a single methodology. However, one has to understand both sides of the coin. On the positive side, hybrid methods carry with them the harmony of two or more chemical or physical disinfectants that contribute together in the inactivation of pathogens commonly present in water. This implies that only a small concentration of the chemical or short exposure to the physical technique is generally required. Interestingly, this translates to less expenditure for the chemical disinfectant or less energy requirement for the physical exposure due to its shorter usage periods. Additionally, hybrid disinfection also gives rise to fewer disinfection by-products; in other words, although DBPs are formed, their concentrations appear to be much less compared to the concentration typically encountered when a single chemical is used. Apparently, one can expect less potential health hazards to humankind. The negative side appears to be the appropriate selection of the combination, which in turn depends on several factors, such as the type and quality of water to be treated, environmental factors, type of pathogens being handled, and end use, to name a few. Although these factors are equally relevant for any single chemical or physical methodology, it becomes strikingly important when several chemicals/physical techniques have to be selected appropriately. As far as microbial inactivation is concerned, the effectiveness, as stated in most of the research findings, is always better than when employing individual methodologies. Myriad combinations have been tried, as discussed in this chapter, and several possibilities are still waiting to be explored. Therefore, research in this arena will continue to allure scientists in order to develop newer technologies that are effective, cost-effective, and ensure wholesome and safe drinking water for humankind.

Questions

1. Explain the principles of disinfection by hybrid techniques.
2. Discuss various factors that influence the efficiency of a hybrid technique.
3. What are advanced oxidation processes? Name some methods based on it.
4. Differentiate between the peroxone process and ozonation.
5. Describe the mechanism of microbial inactivation by the peroxone technique.
6. Explain how a synergistic effect can be brought in water disinfection by the use of hybrid techniques.
7. What is the role of metal ions in any hybrid disinfection method?
8. List all the combinations of a chemical method with a physical method and explain any one in detail.
9. How does ultrasound enhance the chemical disinfection process?
10. Explain the photo-Fenton process.
11. Define photocatalysis. Describe its use as a water disinfection technique.
12. Do hybrid technologies result in disinfection by-products? Explain with suitable examples.

References

Ayyildiz, O., Sanik, S., and Ileri, B. (2011). Effect of ultrasonic pretreatment on chlorine dioxide disinfection efficiency. *Ultrasonics Sonochem.* 18(2):683–688.

Benschoten, J.V., Jensen, J.N., Brady, T.J., Lewis, D.P., Sferrazza, J., and Neuhausser, E.F. (1993). Response of zebra mussel veligers to chemical oxidants. *Water Res.* 27(4):575–582.

Bernbaum, C.M. (1981). Criteria for analyzing interactions between biologically active agents. *Adv. Cancer Res.* 35:269.

Bernbaum, C.M. (1985). The expected effect of a combination of agents: The general solution. *J. Theor. Biol.* 114:413.

Boucher, R.M.G., Pisano, M.A., Tortora, G., and Sawicki, E. (1967). Synergistic effects in sonochemical sterilization. *Appl. Microbiol.* 15(6):1257–1261.

Bougeard, C.M.M., Goslan, E.H., Jefferson, B., and Parsons, S.A. (2010). Comparison of the disinfection by-product formation potential of treated waters exposed to chlorine and monochloramine. *Water Res.* 44(3):729–740.

Burleson, G.R., Murray, T.M., and Pollard, M. (1975). Inactivation of viruses and bacteria by ozone with and without sonication. *Appl. Microbiol.* 29:340–344.

Camper, A.K., and McFeters, G.K. (1979). Chlorine injury and the enumeration of water borne coliform bacteria. *Appl. Environ. Microbiol.* 37:633–641.

Clarke, N., and Knowles. (1982). High purity water by using hydrogen peroxide and ultraviolet radiation. *Effluent Water Treatment J.*, September, 335–341.

Clayton, R. (1992, May). UV-catalysed oxidation in water treatment. *Chem. Engineer* 14:23–26.

Dadjour, M.F., Ogino, C., Matsumura, S., and Shimizu, N. (2005). Kinetics of disinfection of *Escherichia coli* by catalytic ultrasonic irradiation with TiO2. *Biochem. Eng. J.* 25(3):243–248.

Dahi, E. (1976). Physicochemical aspects of disinfection of water by means of ultrasound and ozone. *Water Res.* 10:677–684.

Dillert, R., Siemon, U., and Bahnemann, D. (1998). Photocatalytic disinfection of municipal wastewater. *Chem. Eng. Technol.* 21(4):356–358.

Domek, M.J., LeChavallier, M.W., Cameron, S.C., and McFeters, G.A. (1984). Evidence for the role of copper in the injury process of coliforms bacteria in drinking water. *Appl. Environ. Microbiol.* 48:289–293.

Duckhouse, H., Mason, T.J., Phull, S.S., and Lorimer, J.P. (2004). The effect of sonication on microbial disinfection using hypochlorite. *Ultrasonics Sonochem.* 11(3–4):173–176.

Duguet, J.P., Ferray, C., Mallevialle, J., and Fiessinger, F. (1987). La desinfection par l'ozone: Connaissances des mechanisms et applications pratiques. *Eau Ind. Nuisances* 109:31–34.

Ferguson, D.W., McGuire, M.J., Koch, B., Wolfe, R.L., and Aieta, E.M. (1990). Comparing PEROXONE and ozone for controlling taste and odor compounds, disinfection by-products, and microorganisms. *J. Am. Water Works Assoc.* 82:181–191.

Finch, G.R. (1997). *Control of Cryptosporidium through chemical disinfection: Current state-of-the-art.* Portland, OR: AWWARF Technology Transfer Conference.

Glaze, W.H., and Kang, J.W. (1988). Advanced oxidation processes for treating groundwater contaminated with TCE and PCE: Laboratory Studies. *J. AWWA* 88(5):57–63.

Gyurek, L., Liyanage, L., Belosevic, M., and Finch, G. (1996). Disinfection of *Cryptosporidium parvum* using single and sequential application of ozone and chlorine species. Presented at Proceedings of AWWA Water Quality Technology Conference, Boston.

Harrington, D.K., Van Benschoten, J.E., Jensen, J.N., Lewis, D.P., and Neuhauser, E.F. (1997). Combined use of heat and oxidants for controlling adult zebra mussels. *Water Res.* 31(11):2783–2791.

Hoigné, J., and Bader, H. (1978). Ozonation of water: Kinetics of oxidation of ammonia by ozone and hydroxyl radicals. *Environ. Sci. Technol.* 12(1):79–84.

Ikemizu, K., et al. (1987). Ozonation of organic refractory compounds in water in combination with UV radiation. *J. Chem. Eng. Jpn.* 20(4):369–374.

Ireland, J.S., Klostermann, P., Rice, E.W., and Clarke, R.M. (1993). Inactivation of *Escherichi coli* by titanium dioxide photo-catalytic oxidation. *Appl. Environ. Microbiol.* 59:1668–1670.

Jo, C.H., Dietrich, A.M., and Tanko, J.M. (2011). Simultaneous degradation of disinfection byproducts and earthy-musty odorants by the UV/H2O2 advanced oxidation process. *Water Res.* 45(8):2507–2516.

Jyoti, K.K., and Pandit, A.B. (2003). Hybrid cavitation methods for water disinfection. *Biochem. Eng. J.* 14:9–17.

Katzenelson, E., Kletter, E., Schechter, H., and Shuval, H.I. (1974). Inactivation of viruses and bacteria by ozone. *J. Am. Water Works Assoc.* 66:725–729.

Kim, S.M., and Vogelpohl, A. (1998). Degradation of organic pollutants by the photo-Fenton process. *Chem. Eng. Technol.* 21(2):187–191.

Kouame, Y., and Haas, C.N. (1991). Inactivation of *E. coli* by combined action of free chlorine and monochloramine. *Water Res.* 25(9):1027–1032.

Krasner, S.W., Sclimenti, M.J., and Coffey, B.M. (1993). Testing biologically active filters for removing aldehydes formed during ozonation. *J. Am. Water Works Assoc.* 85(5):62–71.

Kusakabe, K., Aso, S., Hayashi, J., and Isomyra, K. (1990). Decomposition of humic acid and reduction of trihalomethane formation potential in water by ozone with UV irradiation. *Water Res.* 24(6):781–785.

Landeen, L.K., Yahya, M.T., and Gerba, C.P. (1989). Efficacy of copper and silver ions and reduced levels of free chlorine in inactivation of *Legionella pnuemophila*. *Appl. Environ. Microbiol.* 55(12):3045–3050.

Liyanage, L.R.J., Finch, G.R., and Belosevic, M. (1997). Sequential disinfection of *Cryptosporidium parvum* by ozone and chlorine dioxide. *Ozone Sci. Eng.* 19:409–423.

Lykins, B.W., Goodrich, J.A., Koffskey, W.E., and Griese, M.H. (1991). Controlling disinfection byproducts with alternative disinfectants. Presented at Proceedings of AWWA Annual Conference, Philadelphia.

McFeters, G.A., and Singh, A. (1991). Effect of aquatic environmental stress on enteric bacterial pathogens. *J. Appl. Bacteriol.* (Symposium Supplement) 70:115S–120S.

Meunier, L., Canonica, S., and von Gunten, U. (2006). Implications of sequential use of UV and ozone for drinking water quality. *Water Res.* 40(9):1864–1876.

MWDSC AND JMM (Metropolitan Water District of Southern California and James M. Montgomery Consulting Engineers). (1989). *Disinfection byproducts in United States drinking waters*, Vol. I. Cincinnati, OH and Washington, DC: EPA and Association of Metropolitan Water Agencies.

Neytzell-de Wilde, F.G. and Chetty, S. (1990). Evaluation of different methods to produce free radicals for the oxidation of organic molecules in industrial effluents and potable water. Water Research Commission Project 388. February.

Parker, J.F.W., and Smith, H.V. (1993). Destruction of oocysts of *Cryptosporidium parvum* by sand and chlorine. *Water Res.* 27(4):729–731.

Paul, D., and Canter, L.W. (1990). Evaluation of photochemical oxidation technology for remediation of groundwater contaminated with organics. *J. Environ. Sci. and Health*, Part A: *Environ. Sci. Eng. Toxicol.* 25(8).

Pereira, M.A., and Phelps, J.B. (1996). Promotion by dichloroacetic acid and trichloroacetic acid of N-methyl-N-nitrosourea-initiated cancer in the liver of female B6C3F1 mice. *Cancer Lett.* 102(1–2):133–141.

Ramseier, M.K., von Gunten, U., Freihofer, P., and Hammes, F. (2011). Kinetics of membrane damage to high (HNA) and low (LNA) nucleic acid bacterial clusters in drinking water by ozone, chlorine, chlorine dioxide, monochloramine, ferrate(VI), and permanganate. *Water Res.* 45(3):1490–1500.

Rizzo, L. (2009). Inactivation and injury of total coliform bacteria after primary disinfection of drinking water by TiO2 photocatalysis. *J. Hazard. Mater.* 165(1–3):48–51.

Sichel, C., Tello, J., de Cara, M., and Fernández-Ibáñez, P. (2007). Effect of UV solar intensity and dose on the photocatalytic disinfection of bacteria and fungi. *Catalysis Today* 129:152–160.

Sierra, G., and Boucher, R.G.M. (1971). Ultrasonic synergistic effects in liquid-phase chemical sterilization. *Appl. Microbiol.* 23(2):160–164.

Smith, H.V., Smith, A.L., and Girdwood, R.W.A. (1989b). The effect of free chlorine on the viability of *C. parvumoocysts*. Publication no. PRU 2023-M. Water Research Centre Marlow, Bucks, UK.

Sökmen, M., Değerli, S., and Aslan, A. (2008). Photocatalytic disinfection of *Giardia intestinalis* and *Acanthamoeba castellani* cysts in water. *Exp. Parasitol.* 119(1):44–48.

Son, Y., Lim, M., and Khim, J. (2009). *Investigation of acoustic cavitation*. New York: Springer-Verlag, pp. 368–379.

Takahashi, N. (1990). Ozonation of several organic compounds having low molecular weight under ultraviolet irradiation. *Ozone Sci. Eng.* 12:1–18.

Tobiason, J.E., Edzwald, J.K., Schneider, O.D., Fox, M.B., and Dunn, H.J. (1992). Pilot study of the effects of ozone and peroxone on in-line direct filtration. *J. AWWA* 84(12):72–84.

Weavers, L.K., and Hoffmann, M.R. (1998). Sonolytic decomposition of ozone in aqueous solution: Mass transfer effects. *Environ. Sci. Technol.* 32(24):3941–3947.

Wolfe, R.L., Stewart, M.H., Liang, S., and McGuiri, M.J. (1989). Disinfection of model indicator organisms in a drinking water pilot plant using peroxone. *Appl. Environ. Microbiol.* 55(9):2230–2241.

Yayha, M.T., Landeen, M.K., Kutz, S.M., and Gerba, C.P. (1989). Swimming pool disinfection: An evaluation of the efficacy of copper:silver ions. *J. Environ. Health* 51:282–285.

5

Cavitation-Based Disinfection Techniques

5.1 Introduction

In the previous chapters, we discussed and assessed different techniques that are used to disinfect water for potable use. We have already weighed the pros and cons of most of the established chemical, physical, and hybrid techniques, and on the basis of that we can conclude that there is a need to explore new techniques for water disinfection. Cavitation is one such technique. This chapter essentially describes the novel and emerging technique of cavitation and its role in disinfecting water alone or in combination with other established techniques.

It is interesting to note that cavitation is known to occur in nature. It occurs in vascular plants that imbibe water through their xylems. When the water tension becomes so great that the dissolved air in the water occupies the entire vessel elements or tracheids, cavitation occurs. Cavitation is also experienced by aquatic life. Marine animals such as dolphins and tuna are restricted in their speed due to the cavitating effect on their tails at very high speeds. Some variety of shrimps, like the pistol and mantis, catagorised as snapping shrimp, use specialised claws to cause cavitation in order to hunt preys.

For centuries, cavitation has been a phenomenon known for its damaging effect on marine propeller blades. At very high speed, the top velocity of propeller blades increases to a level where the pressure drops to such an extent that water begins to get vapourised and bubbles or cavities are formed. A very high magnitude of shock wave is released when these bubbles collapse, which is responsible for its damaging effect on the blades. Although people still continue to view this as an annoyance, recently a lot of research has been focussed on harnessing this energy for useful physical, chemical, and biological applications. In order to appreciate the usefulness of cavitation, it is essential to first understand its fundamentals.

5.2 What Is Cavitation?

Cavitation is defined as the formation and collapse of vapour cavities in a flowing liquid. A vapour cavity can form anywhere in a flowing liquid where the local pressure is reduced to that of the liquid vapour pressure at the temperature of the flowing liquid (Perry, 1973). Cavitation can be broadly classified into two types. Cavitation produced by the propagation of sound waves in a liquid due to pressure variations is called acoustic cavitation, and that produced by pressure variations in a flowing liquid due to the geometry of the system is called hydrodynamic cavitation.

5.2.1 Acoustic Cavitation

Acoustic cavitation is the growth and collapse of bubbles in liquids induced by the propagation of ultrasonic waves that impose a sinusoidal pressure variation on the transmitting medium, alternately decreasing or increasing the local pressures in the medium. Microscopic bubbles form and grow in size during the "rarefaction" half cycle of the sound wave, and implode during the compression half cycle. The bubbles occur in clouds within the solution, although the lifetime of a single bubble is of the order of microseconds, and its radius is on the order of micrometers (Hua and Thompson, 2000). They can bring about extreme variation in the local conditions. Extreme temperature and pressure gradients occur within and surrounding the cavitational bubble during cavitational collapse. Experimentally, a range of temperatures during bubble implosion has been observed: 12,000 to 14,000 K in aqueous solution (Misik et al., 1995), and pressures of several hundreds of atmospheres, like 1000 atm (Seghal et al., 1979), have been reported. The bubble interiors are under such extreme conditions that light is emitted from the cavitating bubbles, a phenomenon known as sonoluminescence.

5.2.1.1 Factors Affecting Acoustic Cavitation

Mason (1991) has described the parameters that affect acoustic cavitation (Table 5.1). Broadly, parameters related to the sound waves (frequency, intensity, etc.) and those related to the medium under study (viscosity, surface tension, temperature, etc.) influence acoustic cavitation. A recent study on water disinfection by ultrasonic cavitation has revealed that power, duration of treatment, that is, number of passes through the ultrasonic reactor, initial bacterial concentration, and volume of sample treated are important parameters contributing to the efficiency of ultrasound. However, high energy demands of ultrasound suggest that it may be economical to apply ultrasound with another disinfection method (Hulsmans et al., 2010).

TABLE 5.1

Factors Affecting Acoustic Cavitation

Serial No.	Factors	Effect	Possible Reasons
1	Increasing frequency	Decreases cavitation	Rarefaction period becomes too small to allow the molecules to be pulled apart in order to generate cavitation.
2	Increasing medium viscosity	Difficult to produce cavitation	The cohesive forces in a viscous liquid are large, and any increasing viscosity leads to an increase in the amount of energy needed to separate the liquid.
3	Lowering medium surface tension	Increases cavitation	Reduction in the cavitation threshold.
4	Low-medium vapour pressure	Cavitation decreases	Higher vapour pressure, as in the case of volatile liquids, facilitates cavitation.
5	Increasing medium temperature	Cavitation increases	Vapour pressure increases, leading to easier generation of cavities. However, at the boiling point of the medium, a large number of cavitation bubbles are generated concurrently. These will act as a barrier to the transmission of sound and dampen the propagation of ultrasonic energy.
6	Presence of dissolved gas	Increases cavitation	Dissolved gas or small bubbles act as nuclei for cavitation.
7	External applied pressure	Increases intensity of cavitation	Collapse of the cavity becomes more violent.
8	Increased sound intensity	Increases cavitation	However, ultrasonic energy input cannot be increased indefinitely due to its decoupling effect and formation of coalescing bubbles that are stable and may dampen the sonochemical effect.

5.2.1.2 Generation of Ultrasound

Transducers generate ultrasound. A transducer is the name of a device that is capable of converting one form of energy to another. Ultrasonic transducers are designed to convert either mechanical or electrical energy into high-frequency sound (Mason, 1991). There are three main types of transducers:

1. *Gas-driven transducers*: These transducers are used to transmit ultrasonic waves into a gaseous medium like air. Although these types of transducers find application in foam breaking and acceleration of drying processes, their use is limited due to high attenuation of sound waves in air, consequently lowering the intensity of ultrasound and requirement of a very high intensity power source like a sonic horn.

2. *Liquid-driven transducers*: This class of transducers finds wide application in the chemical and food industry for homogenisation, emulsification, and to enhance hydrolysis. These transducers behave like propeller blades operating in an inverse manner; that is, conventionally cavitation occurs due to the speedy movement of the blade through water, whereas in the present case the blade is stationary and the liquid is flowing past it at a particular flow rate. This causes the liquid to vibrate at a particular frequency, which when attains the frequency of ultrasound results in cavitation, producing the resultant effect, such as mixing and homogenisation.

3. *Electromechanical transducers*: Two main types of electromechanical transducers are the piezoelectric and the magnetostrictive. The former are used commonly in the bath and the probe type systems.

   a. **Magnetostrictive**: When a ferromagnetic material like nickel or nickel alloy is exposed to a magnetic field, a change in its dimension is obtained. These find application in metal crystallisation and ore extraction and are limited in use due to their high electric power consumption.

   b. **Piezoelectric**: They are most versatile and commonly used in sonicator systems. These transmit ultrasound through the media via two effects. First, when pressure is applied across the large surfaces of a section, a charge is generated on each face equal in magnitude but opposite in sign. This polarity is reversed if the tension is applied across the surfaces. This is also called the direct effect. Second, if the charge is applied to one face of the section and an equal but opposite charge to the other face, the whole section of crystal will either expand or contract depending on the polarity of the applied charges. On applying rapidly reversing charges to a piezoelectric material, fluctuations in dimensions will be produced. This is also called the inverse effect.

5.2.1.3 Sonication Equipment

Frequently used ultrasonic equipment is listed below. They are suitable for small- to medium-scale applications such as laboratory use or a pilot plant-level application.

5.2.1.3.1 The Ultrasonic Bath

This essentially consists of a stainless steel tank that can have different dimensions to suit specific uses. Transducers are attached to its base in order to generate and transmit ultrasound. They are frequently used as ultrasonic cleaning systems in a wide spectrum of applications, such as in scientific labs for cleaning glassware, hospitals for surgical instruments and hypodermic syringes, the optical industry for cleaning spectacle frames and lenses, and for better adherence of plating during electroplating, to name a few. Cavitation produces the cleaning effect by formation and explosion of cavities with high pressure, resulting in a brushing effect on the surfaces to be cleaned. Moreover, cavities being microscopic in size can reach small crevices usually inaccessible by manual cleaning processes. In many instances cavitation may also enhance the cleaning process of a detergent.

5.2.1.3.2 The Probe System

This is the best for use in laboratories, as it generates maximum ultrasound intensity. It essentially consists of a horn made of metal alloy. Here, the ultrasound is introduced directly into the medium by the transducer via the horn using the principle of waveguide. These are routinely used in biotechnological units, the pharmaceutical industry, chemical labs, and research institutes for applications like cell disintegration, homogenisation, particle dispersion, formulations, emulsification, and tissue (plant and animal) processing. An interesting point to note in the use of probes is that the intensity of cavitation is highest around its tip, which in turn leads to intense local shear gradients that are responsible for processes like homogenisation, disintegration, and emulsification. However, the cavitational intensity decreases at positions away from the tip of the probe. Thus, it is very effective on a lab scale to treat small volumes of sample.

Piezoelectric transducers are commonly employed in the ultrasonic bath and probe. Both types of sonication are generally carried out at a fixed frequency in a range of 20 to 500 kHz. For processing large volumes, flow reactors can be used. The basic principle is to subject a part of the liquid to ultrasound as it passes through a loop system that is linked to a reservoir containing the liquid. Ultrasonic flow cell is one such reactor.

5.2.1.3.3 The Ultrasonic Flow Cell

This type of sonicator is most commonly used to treat large volumes. It can generate high intensity of ultrasound. It consists of a sonic horn transducer that is attached to a flow loop system. Compared to the probes that are used for batch purposes, these are more advantageous, as they can generate a very high intensity of ultrasound operating in a continuous mode. Variation in liquid flow rate and cavitational exposure time is another important feature. This in turn can be used to vary the residence time of the liquid within the system, experiencing acoustic cavitation. However, there are two main disadvantages to the flow system:

- Very high power can result in radical formation and also erosion of the vibrating face, resulting in a contamination of the medium.
- For continuous flow, pumping may be required. This could be a problem in the case of viscous or particulate material-containing systems (Mason, 1991).

5.2.2 Hydrodynamic Cavitation

The momentum balance equation predicts that when a fluid is made to pass through a constriction, due to an increase in its velocity, the static pressure downstream drops. If the pressure falls below a critical value, usually below the vapour pressure of the medium of operating temperature, then small bubbles or vapour cavities are formed in the fluid. The condition at which these fine bubbles can be observed is termed cavitation inception (Yan et al., 1988). An increase in the velocity will result in a further drop in pressure and an increase in the cavitation intensity. Generally, pressure recovery takes place further downstream where these cavities collapse, generating a high-magnitude pressure pulse. If the gas content inside the cavity is small enough, the pressure impulse could be very high (several hundreds of bars)— enough to rupture microbial cells causing its destruction.

5.2.2.1 Cavitation Inception

The term *cavitation threshold* is used to describe the minimum conditions necessary to initiate cavitation. It has been estimated that sound pressures of 1 to 8 bar are required to reach the cavitation threshold for water, depending on its temperature and the dissolved gas content (Hughes, 1961). The probability of cavitation occurring in a flow system is predicted by the calculation of the ratio of forces collapsing cavities to those initiating their formation. This ratio is called *cavitation number* (σ) and is defined as

$$\sigma = \frac{P_r - P_V}{\frac{1}{2}PV^2}$$

where P_r is the recovery pressure and P_V is the vapour pressure of the liquid at the operating temperature. ρ and V are the liquid density and the liquid velocity at the orifice constriction, respectively.

Hence, low cavitation numbers imply lesser collapsing forces and greater initiating forces, that is, greater cavitational activity. Cavitation is expected to occur at cavitation numbers below the cavitation inception number σ_i. Typical values of σ_i in the range of 1 to 2.5 for orifice flow in the pipe (Yan et al., 1988) have been reported. Yan and coworkers (1988) have reported that the value of σ_i decreases with a decrease in the orifice diameter or opening size and increases with increased sharpness of the orifice entrance.

5.2.3 Bactericidal Effects of Cavitation

5.2.3.1 Acoustic Cavitation

Ultrasound has a great potential to be used in a number of fields. Among the various applications, one of the important uses of ultrasound is disruption of biological cells. A lot of work has been reported on ultrasonic disintegration of bacteria, viruses, fungi, and animal cells, for the recovery of intracellular product. Cell disruption is an important step in the downstream processing of a bioproduct. It is known that downstream processing accounts for 80% of the total cost in a typical bioprocess, and cell disruption is a rate-limiting step that has a great impact on the subsequent purification in terms of product yield and generation of fines. Among various options available to bring about cell disruption, mechanical methods such as ultrasound can be used to disrupt microbial cells. Some intracellular products are industrial enzymes such as invertase from yeast, and therapeutic enzymes like L-asparginase from *Erwinia* species. Protein aggregates called inclusion bodies, which are routinely formed during high-level expression of proteins in recombination techniques of bacteria, DNA, and RNA extraction in molecular biology techniques, are a few more examples where ultrasound has been exploited.

Disintegration of bacterial cells by ultrasound has proved to be most valuable when limited quantities of bacteria were available and when the possibility of preparing cell-free enzyme extracts had to be explored. Further, since irradiation of bacteria by ultrasound can be carried out under completely aseptic conditions, it becomes possible to analyse pathogenic bacteria with respect to their endotoxins, enzymes, polysaccharides, and other substances that can be extracted from the cells only after their disintegration. The method, however, has two drawbacks. First, the construction of the apparatus is very expensive, and second, not all bacteria can be disintegrated by ultrasound (Stumpf et al., 1946).

5.2.3.1.1 Forms of Cavitation

There are two forms of ultrasonic cavitation: transient and stable cavitation. Transient cavitation connotes a relatively violent bubble collapse in which localised hot spots of high temperature and pressures occur in very short (in the order of microseconds) bursts that are highly localised in the sonicated medium. These bursts may be accompanied by localised shock waves or the generation of highly reactive chemical radical species. In contrast, the much less violent form of cavitation, stable cavitation, is associated with vibrating gaseous bodies. The nature of this form of cavitation consists of a gaseous body that remains spatially stabilised within, and pulsates due to the ultrasonic field. When such volumetric oscillations are established, the liquid-like medium immediately adjacent to the gas bubble flows or forms fluid velocity streams (termed microstreaming).

Microstreaming resulting from stable cavitation has been shown to produce stresses sufficient to disrupt cell membranes (Scherba et al., 1991).

The mechanism proposed is the onset of turbulence, which creates vortices near which there exist shear rates higher than the shear rates throughout the bulk of the liquid. The shear gradients exerted by microstreaming around bubbles (created by ultrasound), at low sound amplitudes, were sufficient to degrade DNA (Norris and Ribbons, 1971). The damage caused by shear stress is thought to depend on the erosion of the outer cell wall polymers, particularly at weakened places, such as division or budding scars.

5.2.3.1.2 Mechanism of Action

The mechanism by which ultrasound inactivates bacteria has not been conclusively established. Various theories have been reported:

1. *Mechanical effects*: According to the mechanism proposed by Doulah and Hammond (1975), an ultrasound results in the formation of eddies. These eddies are of different scales. Those that are larger than the microorganisms cause the cells to move from place to place, and those that are smaller impart motions of different kinds to the cell wall. These phenomena result in a pressure difference across the cell walls. When this exceeds the cell wall strength, the cell breaks. Kinsloe et al. (1954) have reported that the death rate of microorganisms exposed to intense sound varied exponentially. It was observed that temperature had little effect on the rate of killing, and therefore it was concluded that mechanical disruption was the main mechanism of killing. Marr and Cota-Robles (1954) support this theory. They conducted a study on the sonic disruption of *Azotobacter* where they found that the death rate of the microorganisms was exponential, and the mechanism of killing by ultrasound appears to be due to mechanical effects.

2. *Shear stress*: Shock waves resulting from transient cavitation events may give rise to localised shearing stresses (Thacker, 1973). It was also observed that bubble collapse leads to a turbulent flow and, consequently, high-velocity gradients in localised areas, resulting in increased shear stresses that cause the cell to break. Thacker (1973) also proposes that tensile stress is experienced by the cell. He describes that larger budding yeast cells will be exposed to greater tensile stress than smaller nondividing ones.

3. *Physical forces*: The study conducted by Hamre (1949) reveals that the destruction of microorganisms by ultrasonic energy is primarily the result of physical forces rather than chemical ones (such as oxidation). This was true especially for *Saccharomyces*. However, it has also been reported that the lethal effects of ultrasound may not be similar for smaller organisms like viruses.

4. *Free radical attack*: The possible mechanisms by which the cells are rendered inviable during ultrasonic irradiation include free radical attack (von Sonntag, 1986), including hydroxyl radical attack, and physical disruption of the cell membranes. Once the cell membrane is sheared (a physical consequence of bubble implosion), chemical oxidants can enter the cell and attack internal structures, or vital fluids can be released from the cell, and are degraded in solution.

5. *Disagglomeration*: Ultrasonication can facilitate the disagglomeration of microorganism clusters in solution and thus increase the efficacy of other chemical disinfectants (Hua and Thompson, 2000).

6. *Cavitation*: The most generally accepted view is that cavitation in the suspension medium is primarily responsible for the disruption of microorganisms by ultrasound. The cell breakage by ultrasound is a single hit type of phenomenon, such as the occurrence of a cavitation in close proximity to a cell. The number of cavitation events per minute is independent of the volume of the suspension irradiated and is a function only of the power input to the transducer (Davies, 1959).

5.2.3.1.3 Factors Influencing the Germicidal Effects of Ultrasound

Since the bactericidal effect of ultrasound can be attributed to several mechanisms, as described in the preceding section, it becomes imperative that there may be a spectrum of factors that can directly or indirectly influence the efficacy of acoustic cavitation. Table 5.2 reports important factors affecting ultrasound, which are based on the experimental finding of scientists who have investigated the viability of ultrasonic irradiation as a physical-chemical method for disinfection.

5.2.3.1.3.1 Other Applications of Ultrasound

Ultrasonic cavitation has been applied to achieve goals other than disruption of cells, such as:

1. *Dispersion of bacterial aggregates*: Shropshire (1947) used ultrasound as a means of dispersing suspensions of bacteria. He believed that dispersion prior to enumeration would be valuable in the microbiological analysis techniques.

2. *Inactivate bacteria in water and effluent*: Scherba et al. (1991) undertook a quantitative assessment of the germicidal efficacy of ultrasonic energy. Ultrasonic energy at a frequency of 26 kHz was used to expose aqueous suspensions of bacteria (*E. coli, S. aureus, B. subtilis,* and *P. aeruginosa*), fungi, and viruses. Gram-positive organisms have a thicker and a more tightly adherent layer of peptidoglycan than gram-negative bacteria (*E. coli* and *P. aeruginosa*). Scherba et al. (1991) also sought to determine whether the nature of the cell wall was a

TABLE 5.2

Factors Affecting Germicidal Effects of Ultrasound

Serial No.	Factor	Effect	Reference
Physical Factors			
1	Intensity of ultrasound	Higher intensities enhance inactivation rates. However, for most processes, the increase in process rates does not continue indefinitely with increasing sound intensities.	Hua et al. (1995a), Mason (1991), Hua and Thompson (2000)
2	Dissolved gas	Moderate influence. Mathematical modeling studies suggest presence of active cavitating zone (ACZ) where maximum cavitational effects are seen.	Hua and Hoffmann (1997), Hua et al. (1995b), Mason (1991), Hua and Thompson (2000), Mahulkar et al. (2008)
3	Frequency	Frequencies above 200 kHz enhance the sonochemical reaction rate constants.	Hua and Hoffmann (1997), Petrier et al. (1992, 1996)
Factors Relating to the Microbial Cells			
1	Size	Ultrasonic disintegration of cells decreases with increasing cell size. Larger cells, such as the dividing cells, experience greater tensile stresses than the smaller cells.	Thacker (1973)
2	Shape	Larger, more elongated bacteria are more susceptible to ultrasound than smaller, more compact bacteria.	Thacker (1973)
3	Stage of development	Dividing and diploid cells are more susceptible to death than cells in the stationary phase.	Thacker (1973)
4	Species	Certain species (e.g., certain *Pseudomonas* sp.) exhibit resistance to ultrasonic treatment.	Scherba et al. (1991)

differentiating factor in the destruction of the bacteria. They concluded that this morphological feature was not a factor determining the sensitivity of the bacteria to ultrasound. It was suggested that the target of ultrasonic damage may be the inner cytoplasmic membrane, which consists of a lipoprotein bilayer. This work suggests that ultrasound in the low-kilohertz frequency range has some efficacy in inactivating some disease-causing microbes present in water. Sonocatalytic disinfection uses ultrasonic waves in the presence of titanium dioxide. This is known to release powerful hydroxyl radicals that are effective in killing microbes (Shimizu et al., 2010).

5.2.3.2 *Hydrodynamic Cavitation*

Hydrodynamic cavitation is known to produce the phenomenon of transient cavitation and the associated forces: pressure fluctuations, shock waves, stresses, and temperatures. Based on this, the potential for cell disruption similar to that achieved by ultrasonic cavitation exists. Unlike ultrasonic cavitation, hydrodynamic cavitation has been used to a lesser extent as a means of bacterial inactivation. A decade back there was very little literature that dealt specifically with the effect of hydrodynamic cavitation on bacterial cell viability or the use of hydrodynamic cavitation in water treatment, though it was tried out for wastewater treatment as described in the patented hybrid process of Cavox®. This process used hydrodynamic cavitation, UV radiation, and hydrogen peroxide to oxidise organic compounds present in water at mg/l levels or less. The organic contaminants present in water are oxidised by the hydroxyl and hydroperoxyl radicals, produced by hydrodynamic cavitation, UV radiation, and hydrogen peroxide. Subsequently, the organic compounds are broken down into carbon dioxide, water, halides, and in some cases, organic acids.

Most of the initial studies involving hydrodynamic cavitation were carried out on yeast. Yeast (*Saccharomyces cerevisiae*) is one of the most easily available and cheap sources of microbial biomass. Previous studies (Edebo, 1969) have shown that yeast is one of the most difficult-to-disrupt microorganisms; hence, a mechanical method applicable to yeast should be able to disrupt other organisms. It has often been used for characterising cell disruption equipment and comparing cell lysis techniques (Hetherington et al., 1971; Wiseman et al., 1987; Save et al., 1994; Doulah and Hammond, 1975). This is because it is easily and cheaply available, has generally regarded as safe (GRAS) status, and shows good reproducibility. Disruption studies have been carried out on *Saccharomyces cerevisiae* by Save et al. (1997), Sarkari (1995), and Gopalkrishnan (1996), and cavitation has been found to be one of the main mechanisms of disruption in various equipment (Shirgaonkar et al., 1998). Apart from yeast, hydrodynamic cavitation has also been used to study the disruption of bacterial cells. Hydrodynamic cavitation has been used successfully by Harrison (1990) to disrupt the bacterium *Alcaligenes eutrophus.*

There has been a lot of research happening where hydrodynamic cavitation has been studied as a substitute for ultrasonication, especially in process intensification. Research on the use of hydrodynamic cavitation for water disinfection is slowly gaining momentum. The authors were among the first to explore this area and found that hydrodynamic cavitation is very effective and economical when used in combination with other established chemical disinfectants (Jyoti and Pandit, 2001, 2003a, 2003b, 2004a,b, 2010). A detailed description of the processes involved as well as the experimental work carried out by the authors is described. The possible techniques discussed here are on lab and pilot scale and show considerable promise as future applicable techniques.

5.3 Water Disinfection by Cavitation

This section describes the novel work carried out by the authors, which involved studies on the effect of cavitation on water disinfection for potable use. The main object was to study the effect of cavitation alone and when combined with established conventional techniques, such as treatment with hydrogen peroxide and ozone. Groundwater (bore well water) from the campus of the Institute of Chemical Technology (formerly UDCT), Mumbai was used as source water. Prior to experimentation the microbial population in bore well water was enumerated regularly to ascertain its seasonal variations. This included microbial analysis of heterotropic plate count bacteria and indicator organisms such as total coliforms, feacal coliforms, and feacal streptococci, which have to be monitored, and their number, generally denoted as CFU/ml or total coliforms/100 ml, should fall within the prescribed limits for drinking water laid down by water authorities from time to time. Enumeration of bacteria was done by plate count method as recommended by the American Public Health Association (1985). The flowchart of microbial analysis for heterotropic plate count bacteria and indicator microorganisms are shown in Figure 5.1a and b, respectively.

Preliminary experiments were performed in equipment that has been previously used for microbial cell disruption purposes. These included the high-speed homogeniser, high-pressure homogeniser, ultrasonic horn, ultrasonic bath, and a hydrodynamic cavitation setup. All this equipment is known to generate cavitating conditions. An attempt was made to investigate the effectiveness of each for the disinfection of bore well water for potable use. The experiments were classified into three parts as presented in Figure 5.2.

5.3.1 Experiments with Cavitation

Preliminary experiments were performed on high-pressure and high-speed homogenisers, which are routinely used for cell disruption. Promising results led to further experimentation on acoustic and hydrodynamic cavitation.

1. *High-speed homogeniser (HSH)*: The setup used was a simple rotor stator assembly. A small amount of bore well water was subjected to high-speed homoginisation for a fixed time period. Microbial analysis of samples withdrawn intermittently revealed the extent of disinfection achieved. With an increase in time of treatment, the microbial count was found to decrease due to its higher probability of interaction with cavities formed in the rotor stator. However, the overall rate of disinfection decreased, which may be explained as follows: at the start, the cavities formed and microbes are equal in number, but as microbes get killed, their count reduces, although the number of cavities remains constant, thereby reducing the rate

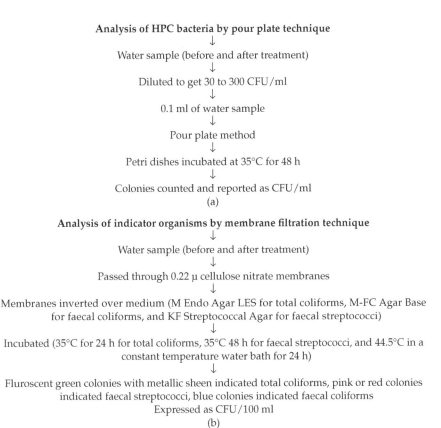

FIGURE 5.1
Microbial assessments of disinfection.

of disinfection. Yet another interesting trend observed was the effect of rpm, which was altered from 1000 to 12,000 during experimentation. A peak was observed at 8000 rpm, at which the rate of disinfection was maximum, after which there was a gradual decline that was attributed to a decrease in residence time of microbes at higher speeds. These exciting results, which were consistent with previous studies on the same setup for microbial cell disruption (Kumar and Pandit, 1999), led to further experiments with a high-pressure homogeniser.

2. *High-pressure homogeniser (HPH)*: Bore well water was subjected to high-pressure homogenisation in a setup that essentially consisted of a positive displacement reciprocating pump, valve, and valve seat assembly. It was a two-stage homogeniser; that is, the material that comes out of the first stage, which could be operated up to 1000 psig, passes to the second stage to undergo further treatment at higher

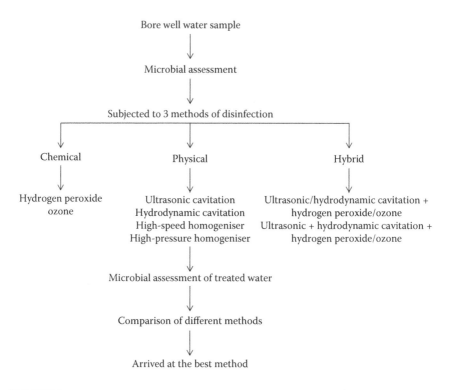

FIGURE 5.2
Disinfection studies.

pressures up to 12,000 psig. An added feature was the provision for recirculation of the homogenised material into the feed tank, if desired. This meant that the material could be subjected to a number of passes, and the time taken to complete one pass was ascertained by finding the flow rate. Experiments conducted were similar to those for the high-speed homogeniser. Similar trends were observed for treatment time and overall rate of disinfection.

In the case of the HPH and HSH a maximum with respect to disinfection rate was observed. This was because as the speed/pressure was increased beyond a certain value, the intensity of cavitation also increased; however, due to the corresponding increase in the velocity, the residence time (time spent by the microorganism in the cavitational zone) decreases, and hence microbes may be only partially affected (paralysed) with respect to their viability, leading to the decrease in the overall destruction rate. In addition to this, when the speed/pressure is increased beyond a certain value, supercavitation occurs and the cavities generated do not collapse and produce the required pressure pulse to destroy the microbes, and thus there is a decrease in the destruction rates. Similar observations have been reported with cell disruption studies (Shirgaonkar et al., 1998). Promising results obtained with initial work on high-pressure and high-speed homogenisers steered the investigation of cavitation for water disinfection.

3. *Ultrasonication*: Acoustic cavitation is induced by passing sound waves into the medium to be treated. This was achieved by subjecting bore well water to acoustic cavitation in an ultrasonic bath and horn. The bath was essentially a stainless steel chamber with piezoelectric transducers at its base to generate ultrasound capable of handling a few litres of sample, whereas the horn was a probe type instrument that had the transducer attached to its tip and generated ultrasound to a small volume of sample placed under the tip (Figure 5.3). Results obtained for treatment time and overall rate of disinfection were very similar to experiments with high-pressure and high-speed homogenisers, and the explanation holds good here also. Temperature rise was prevented by using an ice bath for the probe system, and an intermittent quiet period, when there was no sonication, was used for the bath system.

4. *Hydrodynamic cavitation*: The hydrodynamic cavitation setup used by the authors basically entails a closed-loop circuit consisting of a holding tank, a centrifugal pump, and operational valves (Figure 5.4). The holding tank, 390 mm in diameter and 610 mm in height, had a capacity of 80 l with two baffles 50 mm in width for efficient mixing. The power consumption of the pump supplied by Calama Industries Ltd., India, was 5.5 kW, and a speed of 2900 rpm. The valves used were ball type made of stainless steel. The suction side of the pump is connected to the bottom of the tank. The discharge from the pump branches into two lines, the mainline

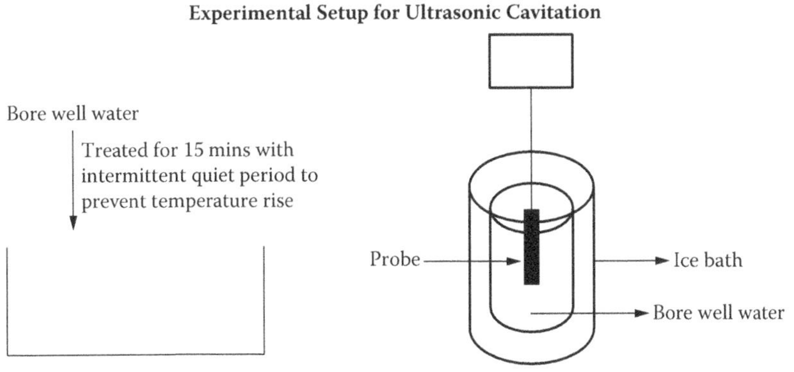

Experimental Setup for Ultrasonic Cavitation

Bore well water

Treated for 15 mins with intermittent quiet period to prevent temperature rise

Probe — Ice bath

Bore well water

FIGURE 5.3
Ultrasonic baths.

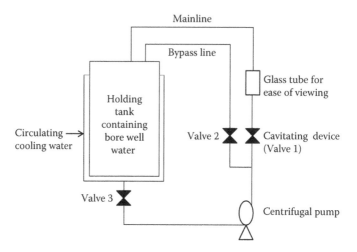

FIGURE 5.4
Hydrodynamic cavitation setup.

which contains the cavitating element, valve 1, which is used to throttle the flow of water, and a bypass line containing valves 2 and 3 at appropriate positions to control the flow of water through the mainline. For the actual experiment, the bypass valve was left open until the pump reached its maximum speed, and then partially or totally closed. The second valve was then throttled to obtain the required pump discharge pressure. The setup could be operated at different discharge pressures. The cooling water circulated through the jacket surrounding the holding tank maintained ambient water temperature. Experiments were also carried out in the presence of a multiple-hole orifice plate placed along the flow of liquid. The orifice plate had 33 holes of 1 mm diameter. The effective flow area was 25.92 mm². This orifice plate changed the cavitating conditions, as has been described elsewhere (Kumar et al., 2000; Vichare et al., 2000). For hybrid experiments involving acoustic cavitation, an ultrasonic flow cell was installed in the hydrodynamic cavitation setup on the discharge side of the pump such that the water, after undergoing hydrodynamic cavitation, is subjected to acoustic cavitation in the flow cell. Experiments were conducted with 75 l of bore well water at pump discharge pressures of 1.72, 3.44, and 5.17 bar with and without multiple-hole orifice plates. Samples were withdrawn at regular time intervals from the top of the tank, which was well mixed due to the returning liquid jet. Trends observed during assessment of various parameters are reported in Table 5.3.

TABLE 5.3

Effects and Possible Explanation of Various Operational Parameters in the
Hydrodynamic Cavitation Setup

Serial No.	Parameter Studied	Effect Observed	Possible Explanation
1	Treatment time	Increasing treatment time increases the number of bacteria killed. Highest percent disinfection was observed after initial 15 min of treatment.	Increasing the time of exposure increases the probability of a cell coming into contact with a collapsing cavity, which would lyse it. When the number of microbes is less than or greater than the number of cavities, the chances of complete disinfection of a given volume of water are expected to be low, and hence each equipment will have an optimum range of operating parameters depending on the microbial count to obtain maximum disinfection (Save et al., 1994; Jyoti and Pandit, 2001).
2	Discharge pressure	Increase in pressure increases percent disinfection and decreases the time required for achieving a certain degree of disinfection.	At higher pressures the number and intensity of cavitation events are higher and the cavitation threshold is reached earlier.
3	Presence of multiple-hole orifice plate	Increased disinfection rate (around twice) compared to that obtained without orifice plate.	Cavitation inception occurs at the fluid shear layer. When water flows out of the orifice plate having multiple holes, the number of shear layers corresponding to each issuing liquid jet (generated from each hole) is formed, and hence the number of cavitational events significantly increases (Vichare et al., 2000). The increase in the yield has been attributed to increased cavitational events, increasing the probability of interaction between the collapsing cavity and the microbe.

5.3.2 Comparison of Cavitation Equipment

After having studied the trends observed for water disinfection in various
equipment discussed in the preceding sections, comparison of these equipment in terms of their efficacy is discussed here. To begin, what is efficacy?
Efficacy may be defined as the ratio of the overall rate of disinfection obtained

TABLE 5.4

Efficacy of Cavitation Methods Based on HPC Bacterial Count

Serial No.	Cavitation Methods	Rate of Energy Dissipation (J/s)	Overall Rate of Disinfection at the End of 15 min of Treatment (CFU killed/s)	Efficacy (CFU killed/J)
1	High-speed homogenisation	45	3794	84
2	High-pressure homogenisation	1137	7537	7
3	Ultrasonic bath	47	9469	203
4	Ultrasonic horn	7	402	55
5	Hydrodynamic cavitation at 1.72 bar	247	117,825	477
6	Hydrodynamic cavitation with multiple-hole orifice plate at 1.72 bar	247	256,425	1038

for a particular technique in a specified time period and the specific rate of energy dissipation of that disinfection technique. Since all of the equipment studied show a marked change in disinfection rate within the first 15 min of treatment, a 15 min time period was considered for comparison. Thus, efficacy was calculated as follows:

Efficacy (CFU killed/J) = overall rate of disinfection at the end of 15 min of treatment (CFU killed/s)/rate of energy dissipation (J/s)

Table 5.4 compares the efficacy of the different cavitational techniques studied by the authors.

Even though higher overall disinfection rates are obtained in the HPH, its energy dissipation is substantially more than the HSH, thereby rendering it less efficacious. The HSH has several slots in the rotor stator assembly where cavitational events can occur, compared to only one slot in the HPH (valve and valve seat). This difference in geometry results in a higher number of microbes to be killed per joule of energy consumed in the HSH. When the volume of water to be treated is small, the efficiency will be better because the entire amount of water will get a chance to pass through the slots where cavitation occurs. When large volumes are to be treated, the liquid at the vicinity of the rotor/stator assembly will be subjected to cavitation, and the entire bulk of the liquid may not experience the cavitational effects. For handling larger volumes the HPH will be better, as the liquid will get a chance to be disinfected as it recirculates through the valve and valve seat with an increasing number of passes through the HPH. Overall, the disinfection rate of the ultrasonic bath was around 23 times more than that of the ultrasonic horn, resulting in better efficacy in spite of its higher energy dissipation than the horn. The bath was found suitable to handle larger volumes due to even distribution of cavities in the bulk liquid. The efficacy of hydrodynamic cavitation for the destruction of heterotropic plate count (HPC) bacteria at different discharge pressures is as follows:

$$1.72 \text{ bar} > 5.17 \text{ bar} > 3.44 \text{ bar}$$

The overall rate of disinfection increases with an increase in the discharge pressure due to an increase in the cavitation intensity. When the pressure is being increased, the valve opening becomes smaller due to throttling of the valve, thus increasing the liquid velocity through the same. This decreases the cavitation number, which in turn leads to an increase in the cavitation intensity. The rate of energy dissipation does not follow the same trend. In fact, the rate of energy dissipation is lowest at 1.72 bar. Thus, the efficacy is better when the hydrodynamic cavitation setup is operated at a low discharge pressure of 1.72 bar. Similar observations can be made for hydrodynamic cavitation in the presence of a multiple-hole orifice plate. However, the efficacy obtained is better than operating without the orifice plate due to a significant increase in the number of cavitational events. Orifice plate geometries can be manipulated along with the operating conditions in a hydrodynamic cavitation setup to enhance the cavitation intensity. Hydrodynamic cavitation is reported to be better than acoustic cavitation in this aspect (Kumar et al., 2000). Typical values obtained are given in Table 5.4. Efficacy of the high-speed homogeniser (84 CFU/J) was almost 12 times higher than that of the high-pressure homogeniser (7 CFU/J) for the destruction of HPC bacteria. The efficacy of ultrasonic bath was found to be 203 CFU/J, almost four times higher than that of the ultrasonic horn, which had an efficacy of 55 CFU/J. The efficacy of hydrodynamic cavitation setup was two times higher in the presence of the multiple-hole orifice plate. These results clearly indicate that energy dissipated and overall rate of disinfection are important factors that govern the efficacy of any given technique. Moreover, an efficient technique may not always be suitable for large-scale disinfection as in the case of the high-speed homogeniser.

5.3.3 Experiments with Chemicals

In order to compare the efficiency of synergistic processes, it is necessary to first quantify disinfection by the chemical method alone. Disinfection by chemicals has been discussed in detail in the previous chapters. We have witnessed the emergence of newer chemical techniques that have several advantages over traditional chlorination, although the former continues to remain the method of choice in several countries globally. This part deals with experiments using established chemical disinfectants other than traditional chlorination, such as hydrogen peroxide and ozone.

5.3.3.1 Hydrogen Peroxide

Hydrogen peroxide is a powerful disinfectant on account of its oxidising property. When a known amount of hydrogen peroxide is added to water, highly reactive hydroxyl radicals are produced, which kills the microbes. Different concentrations of hydrogen peroxide, ranging from 50 to 150 mg/l, were added to bore well water, and microbial analysis was conducted at regular intervals. The study revealed that disinfection was very quick and maximum disinfection was obtained at the end of 15 min of treatment, and higher concentrations required less time for treatment. This effect may be because initially the concentrations of both H_2O_2 and CFU/ml are high. The interaction levels are also high. As the disinfection process proceeds, with time, even though the CFU/ml reduces, hydrogen peroxide is consumed due to dissociation, which in turn slows down the disinfection rate. An interesting observation made was that the hydrogen peroxide did not decompose significantly during the experimental time period, as found out by carrying out iodometric titration using sodium thiosulphate. This confirmed that it was available for disinfection of water.

5.3.3.2 Ozone

Ozone was selected as the second chemical disinfectant by the authors due to its wide use for water treatment globally. It was generated in an ozone generator, which worked on the principle of corona discharge (Jyoti and Pandit, 2010). Air that is used as the feed gas in the generator produces ozone by cleavage of the double bond of an oxygen molecule that produces two short-lived oxygen atoms; these atoms react rapidly with oxygen molecules to produce ozone (Brink et al., 1991; Glaze, 1987). Ozone concentration was then determined by titration with sodium thiosulphate by the iodometric titration procedure, as described in the *Standard Methods* (1985). A stock solution of ozone of a known concentration was prepared by passing ozone into a fixed amount of sterile distilled water for a specified period. An appropriate amount of this ozone solution corresponding to 0.5 to 4 mg/l ozone was used as the dosage for various experiments. In order to ascertain that ozone was available for disinfection, aqueous ozone concentration was determined using an iodometric method (Eaton et al., 1995). This method has been reported in the literature for ozone determination (Caprio et al., 1982; Singer and Zilli, 1975; Smith et al., 1992).

The iodometric method is based on the liberation of free iodine from potassium iodide (KI) solutions by reaction with ozone (Eaton et al., 1995). Gradual decomposition was observed over a period of time, and around 60%

was degraded at the end of 15 min of treatment. Results obtained with ozone were similar to that obtained with hydrogen peroxide; that is, the rate of disinfection was rapid in the initial 15 min of treatment, and higher concentrations required less time for disinfection to occur.

5.3.4 Comparison of Chemical Techniques

The Chick–Watson model (Hassen et al., 2000), which is most widely used to determine microbial inactivation by disinfectants, was used to determine the rate and order of reaction (disinfection). The data obtained for chemical disinfection with hydrogen peroxide and ozone were fitted into the correlation given below and typical results are reported in Table 5.5.

The Chick–Watson model is the most widely used tool to determine microbial inactivation by disinfectants.

$$Ln\ (N/N_0) = -kC^n t$$

where:

N/N_0 = Ratio of number of organisms at time t to time zero

C = CONCENTRATION of disinfectant (which must be constant)

n = Empirical constant or exponent

K = Death rate

N_0 = Initial number of organisms

N = Surviving number of organisms

t = Duration of sterilisation

TABLE 5.5

Rate Constant and Exponent Obtained for Chemical Disinfection after Fitting Data into the Chick–Watson Correlation

Concentration Range: 50 to 150 mg/l Hydrogen Peroxide for HPC Bacteria and 5 to 40 mg/l for Indicator Microorganisms					
Serial No.	Chemical Disinfectant	Concentration Range	Type of Microorganism	Rate Constant, k (mg l^{-1} min^{-1})	Exponent, n
1	Hydrogen peroxide	50 to 150 mg/l	HPC bacteria	2.53×10^{-4}	0.94
		5 to 40 mg/l	Total coliforms	2.37×10^{-3}	0.30
			Faecal coliforms	3.22×10^{-3}	0.17
			Faecal streptococci	2.64×10^{-3}	0.25
2	Ozone	0.5 to 4 mg/l	HPC bacteria	4.29×10^{-2}	0.85
			Total coliforms	6.16×10^{-2}	1
			Faecal coliforms	7.70×10^{-2}	1.33
			Faecal streptococci	5.86×10^{-2}	1.27

If the logarithm of the number of survivors is plotted as a function of time of exposure, a straight line should be obtained where the negative slope defines the death rate. The death rate alone only indicates the surviving fraction of the initial population over the treatment period.

The underlying hypothesis of disinfection is (Langlais et al., 1991):

1. The disinfectant concentration, C, is constant during the reaction time, t.

2. The microorganisms must be a single strain of synchronous development.

3. The killing action must be of a single-hit and single-site type.

The death rate k reflects the speed with which the disinfection takes place, and the exponent n reflects the dependency on C or the usefulness of the oxidant, that is, the concentration of the oxidant. For both hydrogen peroxide and ozone the death rate was found to increase with concentration of oxidant and contact time. The death rate was found to be independent of the initial concentration of microorganisms, but it depends strongly on the type of microorganism to be disinfected. In the study, the k value obtained for the indicator microorganisms was approximately 10 times more than that obtained for the HPC bacteria, and among the indicator microorganisms, the death rate for faecal coliforms was higher than that for the other microbes. This difference in the death rate can be attributed to the difference in the cell wall structures of the microorganisms due to which the susceptibility of the microbes to the chemical oxidant differs. The HPC bacteria were found to be more resistant than the indicator microorganisms. Similar observations were made by Wolfe et al. (1989). The dependency C was highest for the HPC bacteria and least for faecal coliforms. Here again, the type of bacteria and their cell wall structure play an important role, which affects their susceptibility to the oxidant. The HPC bacteria, being most resistant among the various microbes studied here, showed a strong dependence on the concentration of the oxidant used. Higher concentrations of the oxidant will result in better disinfection, and therefore the concentration becomes very critical. However, in the case of the faecal coliforms, the disinfection will be faster due to the higher sensitivity of the microbe to the oxidant. If oxidant concentration is not critical then any amount can be added irrespective of the concentration of molecules.

The values obtained for death rate were much higher (approximately 10 to 20 times) in the case of ozone than those obtained in the case of hydrogen peroxide. It is generally accepted that molecular ozone is a more effective biocide than hydroxyl radicals, since the latter are very short-lived and nonselective species (Duduet et al., 1987). Moreover, Ishizaki et al. (1987) found that ozone penetrates the cell membrane and reacts with cytoplasmic substances. In addition, the chromosomal deoxyribonucleic acids may be degraded, being one of the factors responsible for the cell killing. The n value for ozone is also much higher than that obtained for hydrogen peroxide, which means that there is a greater dependency on C (concentration of the oxidant) for ozone. This may be attributed to the fact that ozone is unstable and undergoes rapid decomposition. Therefore, the availability of the oxidant for the inactivation of the microbes is very critical. The decomposition studies conducted revealed that the decomposition of hydrogen peroxide is not a critical factor, where the hydrogen peroxide concentration practically remained constant over the treatment period.

5.3.5 Hybrid Techniques

Hybrid techniques often result in enhanced disinfection compared to an established single method. They also have an added advantage of fewer disinfection by-products being formed compared to a typical chemical technique such as chlorination. Principles, mode of action, and application, along with merits and demerits of several hybrid processes, have been elaborated separately in Chapter 4. Based on these findings, the authors have investigated some hybrid methods for bore well water disinfection. The approach was to combine cavitation with chemical disinfectants. Three combinations were tried:

1. Two cavitational methods (acoustic and hydrodynamic cavitation)
2. One cavitational method (acoustic or hydrodynamic cavitation) + a chemical disinfectant
3. Two cavitational methods + chemical disinfectant

Initially, a combination of two cavitation methods was studied, acoustic cavitation and hydrodynamic cavitation, in order to obtain higher cavitational intensity and a higher rate of disinfection. An ultrasonic flow cell was installed in the hydrodynamic cavitation setup on the discharge side of the pump such that the water, after undergoing hydrodynamic cavitation, is subjected to acoustic cavitation in the flow cell. The experiments were conducted in a manner identical to only hydrodynamic cavitation setup. In the next attempt, any one cavitation method, either ultrasound or hydrodynamic cavitation, was combined with either hydrogen peroxide or ozone. Finally, a combination of acoustic cavitation, hydrodynamic cavitation, and chemical disinfectant was tried. Various permutations and combinations were tried as shown in Table 5.6 through Table 5.8.

Interesting results were obtained for hybrid experiments, the salient points of which are described below:

- Percentage kill and disinfection rate: Irrespective of the type of hybrid technique performed, that is, two cavitation methods or cavitation and a chemical disinfectant, hybrid methods yielded a higher percentage disinfection as well as an overall rate of disinfection,

TABLE 5.6

Effect of Combining Two Cavitation Methods

Parameter	Disinfection Method	HPC Bacteria	Total Coliforms	Faecal Coliforms	Faecal Streptococci
Percentage disinfection (% killed) at the end of 15 min treatment	US horn	50	55	89	80
	US bath	57	75	47	50
	Hydrodynamic cavitation (5.17 bar) 10 l setup	56	66	57	40
	Hydrodynamic cavitation (1.72 bar) + US flow cell (40 kHz)	99.6	85	60	57
Overall rate of disinfection (CFU killed/s) at the end of 15 min treatment	US horn	402	0.13	0.02	0.02
	US bath	9461	3.85	1.01	0.92
	HC (5.17 bar) without multiple-hole orifice plate	121,950	23.02	12.66	25.45
	Hydrodynamic cavitation (1.72 bar) + US flow cell (40 kHz)	1,786,200	50.56	21.21	35.29

TABLE 5.7

Combination of a Cavitation Method + A Chemical Disinfectant

Parameter	Disinfection Method	HPC Bacteria	Total Coliforms	Faecal Coliforms	Faecal Streptococci
Percentage disinfection (% killed) at the end of 15 min treatment	US bath	57	75	47	50
	US bath + 0.5 mg/l O_3	95	99	95.5	94.2
	Hydrodynamic cavitation (5.17 bar) 10 l setup	56	66	57	40
	Hydrodynamic cavitation (5.17 bar) 10 l setup + 2 mg/l O^3	66	80	88	74
Overall rate of disinfection (CFU killed/s) at the end of 15 min treatment	1 mg/l O_3	1508	0.83	0.39	0.55
	2 mg/l O_3	2121	0.75	0.46	0.66
	3 mg/l O_3	3345	1.85	1.08	1.42
	US horn + 1 mg/l O_3	2209	1.51	0.72	0.75

TABLE 5.8

Combination of Acoustic + Hydrodynamic Cavitation + Chemical Disinfectant

Parameter	Disinfection Method	HPC Bacteria	Total Coliforms	Faecal Coliforms	Faecal Streptococci
Percentage disinfection (% killed) at the end of 15 min treatment	Hydrodynamic cavitation (1.72 bar) + US flow cell (40 kHz)	99.6	85	60	57
	Hydrodynamic cavitation (1.72 bar) + US flow cell (40 kHz) + 5 mg/l H_2O_2	99.7	92	75	70
	Hydrodynamic cavitation (5.17 bar) 10 l setup	56	66	57	40
	Hydrodynamic cavitation (5.17 bar) 10 l setup + 2 mg/l O^3	66	80	88	74
Overall rate of disinfection (CFU killed/s) at the end of 15 min treatment	Hydrodynamic cavitation (1.72 bar) + US flow cell (40 kHz)	1,786,200	50.56	21.21	35.29
	Hydrodynamic cavitation (1.72 bar) + US flow cell (40 kHz) + 5 mg/l H_2O_2	1,798,000	70.10	28.87	48.24
	Hydrodynamic cavitation (5.17 bar) 10 l setup	52,815	4.88	2.82	3.63
	Hydrodynamic cavitation (5.17 bar) 10 l setup + 2 mg/l O^3	65,664	10.72	6.59	6.63

compared to the individual methods. For example, at the end of 15 min of treatment, 99% HPC bacteria were killed using this hybrid method, whereas only 56 and 57% disinfections were obtained with the individual methods of hydrodynamic cavitation and acoustic cavitation, respectively. Similarly, the overall rate of disinfection obtained at the end of 15 min of treatment for hybrid methods was much higher (50 total coliforms/s) than the individual methods of ultrasonication (3 total coliforms/s) and hydrodynamic cavitation (23 total coliforms/s). This effect could be attributed to the synergism created when two cavitational techniques come together. In

other words, hybrid techniques lead to higher cavitational intensity, which is translated in the form of higher percentage and rate
of disinfection than using any one method alone. Numerical simulations of the cavity dynamics equations (Moholkar et al., 1999a,b)
have indicated that it is easier to generate a large number of cavities
hydrodynamically, and it is efficient to collapse them acoustically, as
the acoustic collapse was found to be more violent. This was exactly
confirmed by the author's work. It is now very clear that combining
ultrasound and hydrodynamic cavitation results in improved generation of hydroxyl radicals when the distance between the transducer
and the orifice is 5 to 10 mm (Amin et al., 2010).

- Cavitation-enhanced chemical disinfection efficiency: Cavitation,
  especially that produced acoustically, is known to enhance the efficacy of a chemical disinfectant (Senthilkumar, Sivakumar, and Pandit,
  2000). The literature reports a spectrum of hypotheses and attributes
  this phenomenon to chemical bond rupture of microbial cell membranes, leading to increased permeability to chemical disinfectants,
  increased generation and activities of free radicals when hydrogen
  peroxide or ozone is used (Dahi, 1976), disaggregation of microbial
  flocs (Katzenelson et al., 1974, in Dahi, 1976; Burleson, Murray, and
  Pollard, 1975), or ultrasonic acceleration of diffusion, allowing more
  rapid penetration of the toxic gas molecule into the microorganism
  (Boucher et al., 1967), to name a few. Similar effects, such as those
  discussed here, were clearly observed by the authors in their hybrid
  experiments. For example, percentage reduction of faecal coliforms at
  the end of 15 min of treatment in US bath was 47%. However, addition
  of only 0.5 mg/l O3 substantially enhanced the percentage disinfection to 96%. An interesting observation made was that the quantity of
  chemical disinfectant required to bring about a certain level of disinfection was far less than with methods employing chemicals alone,
  and it depended on the type of microorganism treated. For example,
  in the case of the HPC bacteria, the rate obtained for horn + 1 mg/l
  ozone was similar to the rate obtained with only 2 mg/l ozone. This
  means that, by employing the hybrid process of horn and ozone, the
  concentration of ozone required for disinfection can be reduced by
  half. Similarly, the rate obtained for horn + 1 mg/l ozone was similar
  to the rate obtained with only 3 mg/l ozone in the case of total coliforms and faecal coliforms.
- Enhanced effect of three methods: Since the first two combinations
  resulted in interesting findings, the third methodology of combining acoustic cavitation + hydrodynamic cavitation + ozone/hydro-
  gen peroxide was attempted. As expected, it did result in enhanced
  percentage disinfection and overall rate. Thus, the rate of disinfection increased from 50 total coliforms/s to 70 total coliforms/s when

hydrogen peroxide was combined with hydrodynamic cavitation and acoustic cavitation, which proves that chemical and physical techniques can synergistically work to achieve better disinfection due to an increase in the cavitational intensity when both forms of cavitation are combined together, which probably results in an increase in the decomposition of hydrogen peroxide or ozone, resulting in better efficiencies.

5.3.6 Comparison of All Techniques

5.3.6.1 Efficacy versus Scale of Operation

Various disinfection methods studied by the authors were compared in terms of their efficacy. What is efficacy? It is the ratio of the overall rate of disinfection obtained for a particular technique in a specified time period to the rate of energy dissipation plus the cost of chemicals of that disinfection technique at any specified time, that is, CFU killed/J. The calculations are reported as sample calculations in Chapter 8, Questions 26 and 34. Thermal treatment such as boiling water for drinking is a common method used globally. The efficacy of all the techniques studied in this work is significantly more than the thermal efficacy (Table 5.9).

Thus from Table 5.9, it can be observed that the efficacy of the ultrasonic horn is 55, which is three times higher than the efficacy of thermal treatment. The efficacy of the ultrasonic bath is 203, and the efficacy of the hybrid

TABLE 5.9

Efficacy of Various Techniques as Assessed on HPC Bacteria

Technique	Rate of Energy Dissipation (J/s)	Overall Rate of Disinfection at the End of 15 min of Treatment (CFU killed/s)	Efficacy (CFU killed/J)
Thermal treatment	243.95	5833.33	17.5
Ultrasonic horn	7.31	402	55
Ultrasonic bath	46.63	9469	203
HC with multiple-hole orifice plate (5.17 bar) + H_2O_2	1915	1,046,670	547
Hydrodynamic cavitation (1.72 bar) + US flow cell (40 kHz)	247 + 31.38	1,786,200	6416
Hydrodynamic cavitation (1.72 bar) + US flow cell (40 kHz)+ H_2O_2	247 + 31.38	1,798,000	6459

process of hydrodynamic cavitation + flow cell + hydrogen peroxide is 6459, which is significantly higher (12 and 369 times, respectively) than the efficacy of thermal treatment.

The volume of water treated was different for different methods of disinfection. In order to compare the techniques in terms of their scale of operation, it was necessary to calculate the efficacy as CFU killed/ml.J, which has been reported elsewhere (Jyoti and Pandit, 2010), and a sample calculation is shown at the end of the chapter. The efficacy of the ultrasonic horn was higher (0.55 CFU killed/ml.J) than that of the ultrasonic bath (0.1 CFU killed/ml.J). This clearly points out that if small volumes of water were to be disinfected, then the ultrasonic horn would be ideal. Similarly, in the case of hybrid techniques of ultrasonication and hydrogen peroxide, the ultrasonic horn will be suitable for treating small volumes of water since the efficacy of the horn + hydrogen peroxide (1.41 CFU killed/ml.J) is higher than that of the bath + hydrogen peroxide (0.23 CFU killed/ml.J).

In the case of the hydrodynamic cavitation-based hybrid process, the efficacy of the hydrodynamic cavitation (10 l setup) + ozone is higher (4.8×10^5 CFU killed/ml.J) than the hybrid process of hydrodynamic cavitation (75 litre setup) + hydrogen peroxide (2.9×10^5 CFU killed/ml.J). This again points out that if smaller volumes were to be treated than the 10 l hydrodynamic cavitation setup used and described here, that would be an ideal choice (Table 5.10.)

5.3.6.2 Treatment Cost

For comparison and cost estimation, 100% disinfection criteria were selected since the EEC Guidelines (1975) state that the maximum permissible limit for drinking water is 1000 count of HPC/100 ml at 37°C,

TABLE 5.10

Efficacy of Various Techniques as Assessed on HPC Bacteria

Technique		Rate of Energy Dissipation (J/s)	Rate of Disinfection at the End of 15 min of Treatment (CFU killed/ ml.s)	Efficacy (CFU killed/ml.J)
Ultrasonic horn	HPC bacteria	7.31	4.02	0.55
Ultrasonic bath	HPC bacteria	46.63	4.73	0.10
Ultrasonic horn + H_2O_2	HPC bacteria	7.31	10.33	1.41
Ultrasonic bath + H_2O_2	HPC bacteria	46.63	10.66	0.229
HC (10 l setup) + O_3	HPC bacteria	1373	6.5664	4.8×10^{-3}
HC without multiple-hole orifice plate (5.17 bar) + H_2O_2	HPC bacteria	1915	5.695	2.9×10^{-3}

and the total coliforms/100 ml, fecal coliforms/100 ml, and fecal strepto-cocci/100 ml should be zero. Only the initial rate was considered, as the authors found that the initial 15 min of treatment resulted in maximum disinfections. Table 5.11 gives the cost of the disinfection techniques. The

TABLE 5.11

Disinfection Treatment Cost

Disinfection Technique	Microorganism	Cost (Rs/l)
150 mg/l H_2O_2	HPC bacteria	0.00375
5 mg/l H_2O_2	Total coliforms	0.000125
2 mg/l O_3	HPC bacteria	0.00004
	Total coliforms	0.00004
High-speed homogeniser	HPC bacteria	0.0034
High-pressure homogeniser	HPC bacteria	0.027
Ultrasonic horn	HPC bacteria	0.05
	Total coliforms	1.6
Ultrasonic horn + 150 mg/l H_2O_2	HPC bacteria	0.03[a]
Ultrasonic horn + 5 mg/l H_2O_2	Total coliforms	1.38[a]
Ultrasonic horn + 2 mg/l O_3	HPC bacteria	0.1[a]
	Total coliforms	0.36[a]
Ultrasonic bath	HPC bacteria	0.001
	Total coliforms	0.03
Ultrasonic bath + 150 mg/l H_2O_2	HPC bacteria	0.0007[a]
Ultrasonic bath + 5 mg/l H_2O_2	Total coliforms	0.03[a]
Ultrasonic bath + 2 mg/l O_3	HPC bacteria	0.0013[a]
	Total coliforms	0.04[a]
Hydrodynamic cavitation (1.72 bar)	HPC bacteria	0.002
	Total coliforms	0.44
Hydrodynamic cavitation (5.17 bar)	HPC bacteria	0.0014
	Total coliforms	0.17
Hydrodynamic cavitation (5.17 bar) + 150 mg/l H_2O_2	HPC bacteria	0.0013[a]
Hydrodynamic cavitation (5.17 bar) + 5 mg/l H_2O_2	Total coliforms	0.18[a]
Hydrodynamic cavitation (1.72 bar) + US flow cell (40 kHz)	HPC bacteria	0.0004
	Total coliforms	0.10
Hydrodynamic cavitation (1.72 bar) + US flow cell (40 kHz) + 150 mg/l H_2O_2	HPC bacteria	0.0004[a]
Hydrodynamic cavitation (1.72 bar) + US flow cell (40 kHz) + 5 mg/l H_2O_2	Total coliforms	0.09[a]
Hydrodynamic cavitation 10 l setup (5.17 bar)	Total coliforms	0.28
Hydrodynamic cavitation 10 l setup (5.17 bar) + 2 mg/l O_3	Total coliforms	0.14[a]

[a] Cost of H_2O_2 (approx. Rs 0.000125/–) and cost of ozone (approx. Rs 0.00004/–) used.

method employed to calculate these values has been reported as a sample calculation in Chapter 8.

Cost plays a vital role in the selection of a suitable disinfection technique, which in turn would affect the overall economics of a water treatment scheme. An ideal disinfection technique is one that is able to bring down the bacterial population to the desired level, and is also economical without any harmful by-product formation. Chlorine has been used extensively because it is very cheap. However, use of chlorine also results in carcinogenic by-products like trihalomethanes. This was elaborately discussed in Chapter 2. Previous discussions indicate that hybrid methods like the use of hydrodynamic cavitation, acoustic cavitation, and hydrogen peroxide, and hydrodynamic cavitation and ozone appear to be such techniques. However, the cost of treatment is considerably more than that for the use of hydrogen peroxide or ozone alone (Table 5.11). The higher cost of hybrid methods can be attributed to the high energy requirement. On the other hand, chemical disinfection techniques, that is, treatment with hydrogen peroxide or ozone, was cheaper by one or two orders of magnitude than the hybrid methods described in this chapter. However, the disadvantages associated with chemical treatment, such as the formation of toxic by-products, could be reduced or altogether eliminated by these hybrid methods. Therefore, it appears that bringing down the energy requirement can reduce the cost of treatment. There are a number of situations where the water reservoir and the treatment plants are located at a considerable distance and the water is either pumped from the reservoir or made to flow by gravity. Thus, many times water is available at the water treatment plant at considerable hydrostatic heads or pressures, which are then reduced using an elaborate pressure reduction station to make it suitable for chemical treatment, such as chlorination, ozonation, and so forth. The design of these pressure reduction stations can be changed so as to make them work in a hydrodynamic cavitation mode, without the supply of any additional energy. This is likely to reduce the treatment cost and also the quantity and cost of the chemicals used in the subsequent treatment, as is evident from the reduced use of H_2O_2 (5 mg/l as against 150 mg/l) for the same level of disinfection, as found out in this study. Thus, hydrodynamic cavitation, if used in a hybrid mode, shows considerable promise.

The authors' study revealed some major findings for water disinfection, which are summarised below:

1. Cavitation is a promising technique for water disinfection. Being a non-chemical method, its use does not result in the formation of any toxic by-products, as is the case with chemical treatments using chlorine.

2. It is an energy-efficient process and hence can be considered a potential supplementary technique for a large-scale water treatment scheme.

3. Pretreatment with hydrodynamic cavitation reduces the disinfection chemical consumption substantially by almost 50 to 90%.

4. Hybrid techniques are far superior for treating water than any individual physical treatment technique discussed in this chapter.

5. Hybrid technique not only reduces the HPC bacteria (CFU/ml) but also reduces the total coliforms, faecal coliforms, and faecal streptococci, which are considered the indicators of pollution in drinking water.

5.4 Current Status and Path Ahead

Continuing research on cavitation over the last decade has further corroborated the authors' work. Now, hydrodynamic cavitation is considered one of the potential techniques for water disinfection, especially in the rural areas (Mezule et al., 2009). However, in comparison to the conventional techniques, it appears to be expensive and suitable only in the small-scale treatment of water. This is the major roadblock to industrial-scale applications. It is believed that with research directed toward intensification of cavitation and efficient reactor design this hurdle can be easily overcome in the future (Gogate, 2007). Cavitational reactors are reported to be very effective for water disinfection, and newer reactors have been designed that offer effective design and scale-up possibilities (Kumar and Moholkar, 2007). Study on different chamber designs of hydrodynamic cavitation-based reactors, as well as operating parameters for inactivation of *E. coli* in water disinfection, have revealed that this method holds a lot of promise (Arrojo et al., 2008). Mathematical modeling studies for predicting cavitational intensity in hydrodynamic cavitational reactors in the presence of orifice plates of different geometries have been undertaken for further optimising the process (Sharma et al., 2008). Designing cavitation reactors with a novel cluster approach for cavity bubbles has also been tried successfully (Kanthale et al., 2008). Bactericidal effects by acoustic cavitation have been researched extensively and also discussed in the preceding sections elaborately. Presently, more work is being directed toward bactericidal effects of hydrodynamic cavitation, especially on indicator organisms such as *E. coli*, which was lacking before. It has been demonstrated that physicochemical effects of cavitation are responsible for the destruction of cells and intensity of cavitation, and the microbial load and the dissolved gas content are influencing parameters in disinfection. Studies on water disinfection by hydrodynamic cavitation have extended gradually from its effect on indicator microorganisms to zooplanktons. Recently, cavitation has been reported as a practical tool for seawater disinfection (Sawant et al., 2008). It is continued to be reported to intensify several physical and chemical operations and finds wide application in biotechnology and biochemical engineering, such as microbial cell disruption for release of valuable products, water disinfection, wastewater treatment, emulsification, and molecular biology applications such as gene transfer. An interdisciplinary approach has been recommended to effectively use cavitational reactors (Gogate and Kabadi, 2009). The latest approach for water disinfection has been hybrid techniques involving hydrodynamic cavitation and disinfectants. A novel technique of water disinfection by ozone assisted by a liquid whistle reaction based on hydrodynamic cavitation has been reported. The combination was found to be cost effective compared to individual techniques of hydrodynamic cavitation and ozone for disinfecting the indicator organism, *E. coli* (Chand et al., 2007). Newer reactors based on the use of microwire mesh and a downstream nozzle are being experimented with, against the conventional orifice plate geometries. A hybrid method combining such a cavitational reactor with chlorine dioxide resulted in higher inactivation rates of *E. coli*. Moreover, the concentration of chemical required is generally found to compensate for the higher energy requirement of the cavitation setup (Maslak and Weuster-Botz, 2011).

Patenting is yet another interesting development in the area of water disinfection and cavitation that is becoming rampant. Several water purification systems based on the hydrodynamic cavitation technology have been patented, which could be used for treatment of groundwater, wastewater, and industrial process water, as a stand-alone process or as a part of a treatment train. The DYNAJETS® cavitating jet systems developed by Dynaflow are one such example that is based on an inexpensive hydromechanical source of cavitation. It has been reported to achieve both microbial reduction and organic pollution control with higher efficiencies than ultrasonic devices. Hydrodynamic cavitation has been used by a Las Vegas company, VRTX Technologies, to effectively conserve water in cooling towers. It has been reported to reduce scale formation, corrosion, and bacteria in an energy-efficient manner (www.watersmartinnovations.com). It also resulted in less water consumption and water discharge. Several patent-pending technologies on hydrodynamic cavitation for water disinfection have sprung up recently, such as the Vorsana Radial Counterflow Reactor, which is designed for large-scale treatment of water (http://www.vorsana.net/home.html). The patented design of a water disinfection unit, Waterbotruff 7.5 A, is yet another example of the upcoming relevance of cavitation against chemical disinfection (http://www.dewacor.com/enertech.html). The unit claims to destroy pathogens and organic pollutants in water. HyCa Technologies is another example of an emerging Indian clean-tech company that uses its own patent-pending technology HyCator™ for several physical, chemical, and biological applications. Its outstanding achievements in the domain of effluent treatment, cooling towers, mixing, ballast, and drinking water treatment technologies have been featured in media and earned HyCa Technologies recognition as an upcoming company.

Thus, the future holds a lot of promise for cavitation technology. Acoustic cavitation will continue to enjoy its place as an effective method, especially for small-scale applications. Hydrodynamic cavitation, with further optimisation, is likely to be viewed as a green technology in the water and wastewater industry on a large scale. However, there are some lacunae that need to be addressed. Disinfection by-products formed by cavitational hybrid methods have to be ascertained. The presence of active but not culturable (ABNC) bacteria should also be considered, as it could lead to regrowth of these microorganisms in the distribution system. It is believed that the quest for pure drinking water will continue as human-kind will face acute water shortage in the future. More and more focus of the scientific community globally will be directed toward addressing this issue. As we all know, necessity is the mother of invention, and one can be optimistically reassured that there will definitely be a solution to every problem.

Questions

1. What is cavitation? Describe briefly different types of cavitation.
2. Enlist factors affecting acoustic cavitation. Describe any one in detail.
3. What is the effect of medium viscosity and surface tension on cavitation?
4. What is a transducer? Describe different types of transducers.
5. Describe various sonication equipment and their working principle.
6. Explain the term *cavitation inception*.
7. Give an equation defining cavitation number and explain its significance in hydrodynamic cavitation.
8. How does ultrasound inactivate microorganisms?
9. What factors can influence germicidal effects of ultrasound?
10. How is disinfection of water assessed? Describe in detail enumeration of heterotrophic plate count bacteria as recommended by the American Public Health Association (APHA).
11. What is the effect of rotor speed on disinfection rate in a high-speed homogeniser?
12. What is the effect of increasing pressure on disinfection rate in a high-speed homogeniser?
13. What will be the effect of increasing treatment time on the disinfection rate in an ultrasonicator?
14. With the help of a neat diagram, explain the working of a typical hydrodynamic cavitation setup.

15. Why is the disinfection rate higher in the initial 15 min of treatment in a hydrodynamic cavitation setup?

16. Why is the high-speed homogeniser better than the high-speed homogeniser in terms of efficacy?

17. In terms of efficacy, does the type of sonicator have any role in disinfection?

18. What is the effect of increasing discharge pressures on disinfection rate in a hydrodynamic cavitation setup?

19. Briefly describe the effect of decomposition of ozone and hydrogen peroxide on disinfection.

20. How does change in concentration of a chemical disinfectant like ozone affect the overall disinfection rate?

21. Explain the Chick–Watson model. How can it be used to compare the performance of chemical disinfectants? Also give the significance of the rate constant and exponent.

22. How useful are hybrid techniques for water disinfection?

23. What is the role of cavitation in a synergistic process? Elaborate on the advantages.

24. What is an orifice plate? How does it improve the cavitation intensity in a hydrodynamic cavitation setup?

25. Give the mechanism of microbial destruction by a hybrid process involving cavitation and a chemical disinfectant.

26. What are indicator microorganisms? Explain their role in drinking water treatment efficacy.

27. Compare thermal disinfection with cavitation.

28. Comment on sonication efficiency versus scale of operation.

References

American Public Health Association. (1985). *Standard methods for the examination of water and waste water*, 16th ed. Washington, DC: APHA.

Amin, M.T., and Han, M.Y. (2010). Improvement of solar based rainwater disinfection by using lemon and vinegar as catalysts. *Desalination* 276(1–3):416–424.

Arrojo, S., Benito, Y., and Martínez Tarifa, A. (2008). A parametrical study of disinfection with hydrodynamic cavitation. *Ultrasonics Sonochem.* 15(5):903–908.

Boucher, R.M.G., Pisano, M.A., Tortora, G., and Sawicki, E. (1967). Synergistic effects in sonochemical sterilization. *Appl. Environ. Microbiol.* 15:1257–1261.

Brink, D.R., Langlais, B., and Reckhow, D.A. (1991). Introduction. In B. Langlais, D.A. Reckhow, and D.R. Brink (eds.), *Ozone in water treatment: Application and engineering*. Chelsea, MI: Lewis Publishers, pp. 1–10.

Burleson, G.R., Murray, T.M., and Pollard, M. (1975). Inactivation of viruses and bacteria by ozone, with and without sonication. *Appl. Microbiol.* 29(3):340–344.

Caprio, V., Insola, A., Lignola, P.G., and Volpicilli, G. (1982). A new attempt for the evaluation of the absorption constant of ozone in water. *Chem. Eng. Sci.* 37(1):122–124.

Chand, R., Bremner, D.H., Namkung, K.C., Collier, P.J., and Gogate, P.R. (2007). Water disinfection using the novel approach of ozone and a liquid whistle reactor. *Biochem. Eng. J.* 35(3):357–364.

Council Directive 75/440/EEC of June 16, 1975 concerning the quality required of surface water intended for the abstraction of drinking water in the Member States. O.J. No. L194/26.

Dahi, E. (1976). Physicochemical aspects of disinfection of water by means of ultrasound and ozone. *Water Res.* 10(8):677–684.

Davies, R. Observations on the use of ultrasound waves for the disruption of microorganisms. *J. Bacteriol.* 33:481–493.

Doulah, M.S., and Hammond, T.H. (1975). A hydrodynamic mechanism for the disintegration of *Saccharomyces cerevisiae* in an industrial homogenizer. *Biotech. Bioeng.* 17:845–858.

Duduet, J.P., Ferray, C., Mallevialle, J., and Fiessinger, F. (1987). La desinfection par l'ozone: Connaissances des mechanisms et applications pratiques. *Eau Ind. Nuisances* 109:31–34.

Eaton, A.D., Clesceri, L.S., and Greenberg, A.E. (eds.). (1995). Section 4500-Cl B. Iodometric method I. In *Standard methods for the examination of water and wastewater*, 19th ed. Washington DC: American Public Health Association, pp. 4/38–4/39.

Edebo, L. (1969). Disintegration of cells. In D. Perlman (ed.), *Fermentation advances*. New York: Academic Press, pp. 249–271.

Glaze, W.H. (1987). Drinking-water treatment with ozone. *Environ. Sci. Technol.* 21(3):224–230.

Gogate, P.R. (2007). Application of cavitational reactors for water disinfection: Current status and path forward. *J. Environ. Manage.* 85(4):801–815.

Gogate, P.R., and Kabadi, A.M. (2009). A review of applications of cavitation in biochemical engineering/biotechnology. *Biochem. Eng. J.* 44(1):60–72.

Gopalkrishnan, J. (1996). Studies in cell disruption, M.Sc. (Tech) thesis, University of Bombay.

Hamre D. (1949). The effect of ultrasonic waves upon *Klebsiella pneumoniae*, *Saccharomyces cerevisiae*, *Miyagawanella felis*, and influence virus A. *J. Bacteriol.* 57:279–295.

Harrison, S.T.L. (1990). The extraction and purification of PHB from *Alcaligenes eutorphus*. Ph.D. thesis, University of Cambridge, U.K.

Hassen, A., Mahrouk, M., Ouzari, H., Cherif, M., Boudabous, A., and Damelincourt, J.J. (2000). UV disinfection of treated wastewater in a large-scale pilot plant and inactivation of selected bacteria in a laboratory UV device. *Bioresource Technol.* 74(2):141–150.

Hetherington, P., Follows, M., Dunnill, P., and Lilly, M.D. (1971). Release of protein from baker's yeast (*Saccharomyces cerevisiae*) by disruption in an industrial homogeniser. *Trans. Inst. Chem. Eng.* 49:142–148.

Hua, I., Hochemer, R.H., and Hoffmann, M.R. (1995a). Sonochemical degradation of p-nitrophenol in a parallel plate near-field acoustical processor. *Environ. Sci. Technol.* 29:2790–2796.

Hua, I., Hochemer, R.H., and Hoffmann, M.R. (1995b) Sonolytic hydrolysis of p-nitrophenol acetate: The role of supercritical water. *J. Phys. Chem.* 99:2335–2342.

Hua, I., and Hoffmann, M.R. (1997). Optimization of ultrasonic irradiation as an advanced oxidation technology. *Environ. Sci. Technol.* 31:2237–2243.

Hua, I., and Thompson, J.E. (2000). Inactivation of *Escherichia coli* by sonication at discrete ultrasonic frequencies. *Water Res.* 34(15):3888–3893.

Hughes, D.C. (1961). The disintegration of bacteria and other micro-organisms by the M.S.E. Mullard ultrasonic disintegrator. *J. Biochem. Microbiol. Tech. Eng.* 3:405–433.

Hulsmans, A., Joris, K., Lambert, N., Rediers, H., Declerck, P., Delaedt, Y., Ollevier, F., and Liers, S. (2010). Evaluation of process parameters of ultrasonic treatment of bacterial suspensions in a pilot scale water disinfection system. *Ultrasonics Sonochem.* 17(6):1004–1009.

Ishizaki, K., Sawadaishi, K., Miura, K., and Shinsiki, N. (1987). Effect of ozone on plasmid DNA of *Escherichia coli in situ. Water Res.* 21:823–827.

Jyoti, K.K., and Pandit, A.B. (2001). Water disinfection by acoustic and hydrodynamic cavitation. *Biochem. Eng. J.* 7:201–212.

Jyoti, K.K., and Pandit, A.B. (2003a). Hybrid cavitation methods for water disinfection. *Biochem. Eng. J.* 14:9–17.

Jyoti, K.K., and Pandit, A.B. (2003b). Hybrid cavitation methods for water disinfection: Simultaneous use of chemicals with cavitation. *Ultrasonic Sonochem.* 10:255–264.

Jyoti, K.K., and Pandit, A.B. (2004a). Effect of cavitation on chemical disinfection efficiency. *Water Res.* 38:2249–2258.

Jyoti, K.K., and Pandit, A.B. (2004b). Ozone and cavitation for water disinfection. *Biochem. Eng. J.* 18:9–19.

Jyoti, K.K., and Pandit, A.B. (2010). *Cavitation—A new horizon in water disinfection.* Berlin: VDM Verlag Dr. Muller GmbH & Co. KG.

Kanthale, P., Ashokkumar, M., and Grieser, F. (2008). Somoluminescence, sonochemistry (H2O2 yield) and bubble dynamics: Frequency and power effects. *Ultrasonics Sonochem.* 15:143–150.

Kinsloe, H., Ackerman, E., and Reid, J.J. (1954). Exposure of microorganisms to measured sound fields. *J. Bacteriol.* 68:373–380.

Kumar, K.S., and Moholkar, V.S. (2007). Conceptual design of a novel hydrodynamic cavitation reactor. *Chem. Eng. Sci.* 62(10):2698–2711.

Kumar, S., and Pandit, A.B. (1999). Modelling hydrodynamic cavitation. *Chem. Eng. Technol.* 22:1017–1027.

Kumar, P.S., Sivakumar, M., and Pandit, A.B. (2000). Experimental quantification of chemical effects of hydrodynamic cavitation. *Chem. Eng. Sci.* 55(9):1633.

Langlais, B., Reckhow, D.A., and Brink, D.A. (eds.). (1991). *Ozone in water treatment: Application and engineering.* American Water Works Research Foundation, Chelsea, MI: Lewis Publishers.

Mahulkar, A.V., Bapat, P.S., Pandit, A.B., and Lewis, L.M. (2008). Steam bubble cavitation *AICHE J.* 54(7): 1711–1724.

Marr, A.G., and Cota-Robles, E.H. (1954). Sonic dispersion of *Azotobacter vinelandii. J. Bacteriol.* 74:79–86.

Maslak, D., and Weuster-Botz, D. (2011). Combination of hydrodynamic cavitation and chlorine dioxide for disinfection of water. *Eng. Life Sci.* DOI: 10.1002/elsc.201000103.

Mason, T.J. (1991). *Practical sonochemistry: User's guide to applications in chemistry and chemical engineering.* Chichester: Ellis Horwood.

Mezule, L., Tsyfansky, S., Yakushevich, V., and Juhna, T. (2009). A simple technique for water disinfection with hydrodynamic cavitation: Effect on survival of *Escherichia coli. Desalination* 248(1–3):152–159.

Misik, V., Miyoshi, N., and Riesz, P. (1995). EPR spin-trapping study of the sonolysis of H2O/D2O mixture: Probing the temperatures of cavitation regions. *J. Phys. Chem.* 99:3605–3611.

Moholkar, V.S., Pandit, A.B., and Warmoeskerken, M.M.C.G. (1999a). Characterization and optimization aspects of a sonic reactor. Presented at ICEU-99, New Delhi, India.

Moholkar, V.S., Senthilkumar, P., and Pandit, A.B. (1999b). Hydrodynamic cavitation for sonochemical effects. *Ultrason. Sonochem.* 6:53.

Norris, J.R., and Ribbons, D.W. (eds.). (1971). *Methods in microbiology*, Vol. 5B. London: Academic Press.

Perry, R.H. (1973). *Chemical engineers' handbook*, 5th ed. Tokyo: McGraw-Hill.

Petrier, C., Jeanet, A., Luche, J.L., and Reverdy, G. (1992). Unexpected frequency effects on rate of oxidative processes induced by ultrasound. *J. Am. Chem. Soc.* 114:3148.

Petrier, C., David, B., and Laguian, S. (1996). Ultrasonic degradation at 20 and 500 kHz of atrazine and pentachlorophenol in aqueous solutions: Preliminary results. *Chemosphere* 32:1709.

Sarkari, M. (1995). Process development. M.Sc. (Tech) thesis, University of Bombay.

Save, S.S., Pandit, A.B., and Joshi, J.B. (1997). Use of hydrodynamic cavitation for large-scale microbial cell disruption. *Chem. Eng. Res. Des.* 75(Part C):41–49.

Sawant, S.S., Anil, A.C., Krishnamurthy, V., Gaonkar, C., Kolwalkar, J., Khandeparker, L., Desai, D., Mahulkar, A.V., Ranade, V.V., and Pandit, A.B. (2008). Effect of hydrodynamic cavitation on zooplankton: A tool for disinfection. *Biochem. Eng. J.* 42(3):320–328.

Scherba, G., Weigel, R.M., and O'Brien Jr., W.D. (1991). Quantitative assessment of the germicidal efficacy of ultrasonic energy. *Appl. Environ. Microbiol.* 57(7):2079–2084.

Seghal, C., Steer, R.P., Sutherland, R.G., and Verrall, R.E. (1979). Sonoluminiscence of argon saturated alkali metal salt solutions as a probe of acoustic cavitation. *J. Phys. Chem.* 70:2242–2248.

Senthilkumar, P., Sivakumar, M., and Pandit, A.B. (2000). Experimental quantification of chemical effects of hydrodynamic cavitation. *Chem. Eng. Sci.* 55:1633.

Sharma, R., Tsuchiya, M., and Bartlett, J.D. (2008). Fluoride induces endoplasmic reticulum stress and inhibits protein synthesis and secretion. *Environ. Health Perspect.* 116(9):1142–1146.

Shimizu, N., Ninomiya, K., Ogino, C., and Rahman, M.M. (2010). Potential uses of titanium dioxide in conjunction with ultrasound for improved disinfection. *Biochem. Eng. J.* 48(3):416–423.

Shirgaonkar, I.Z., Lothe, R.R., and Pandit, A.B. (1998). Comments on the mechanism of microbial cell disruption in the high pressure homogeniser. *Biotechnol. Progress.* 14:657–660.

Shropshire, R.F. (1947). Bacterial dispersion by sonic energy. *J. Bacteriol.* 54:325–331.

Singer, P.C., and Zilli, W.B. (1975). Ozonation of ammonia in wastewater. *Water Res.* 9(2):127–134.

Smith, D.W., Mohammed, A., and Finch, G.R. (1992). A bench-scale, two phase reactor of ozone treatment: Feasibility studies for high strength waters. *Ozone Sci. Eng.* 14(5):381–389.

Stumpf, P.K., Green, D.E., and Smith, F.W. (1946). Ultrasonic disintegration as a method of extracting bacterial enzymes. *J. Bacteriol.* 51:487–493.

Thacker, J. (1973). An approach to the mechanism of killing of cells in suspension by ultrasound. *Biochem. Biophys. Acta* 304:240–248.

Vichare, N.P., Gogate, P.R., and Pandit, A.B. (2000). Optimization of hydrodynamic cavitation using model reaction. *Chem. Eng. Technol.* 23:683.

von Sonntag, C. (1986). Disinfection by free-radicals and UV-radiation. *Water Supply* 4:9–10.

Wiseman, A., King, D.J., and Winkler, M.A. (1987). The isolation and purification of protein and peptide products. In D.R. Berry, I. Russell, and G. G. Stewart (eds.), *Yeast biotechnology*. London: Allen & Unwin, pp. 433–464.

Wolfe, R.L., Stewart, M.H., Liang, S., and McGuiri, M.J. (1989). Disinfection of model indicator organisms in a drinking water pilot plant using peroxone. *Appl. Environ. Microbiol.* 55(9):2230–2241.

Yan, Y., Thorpe, R.B., and Pandit, A.B. (1988). Cavitation noise and its suppression by air in orifice flow. In *Proceedings of the International Symposium on Flow Induced Vibration and Noise*. Chicago: American Society of Mechanical Engineers.

6

Novel Disinfection Methods

6.1 Introduction

Microbial disinfection of water can be carried out by various chemical, physical, and hybrid techniques that have been discussed at length in the preceding chapters. It is really interesting to note that although man has experimented with different methodologies for water disinfection, there has never been a universal solution for microbial inactivation for potable waters. The quest for better, efficient, and economical methods continues to intrigue humankind, and research in this arena is augmenting day by day. Globally, scientists are striving to come up with new methods for disinfection of water. In order to throw some light on the emerging and novel techniques of water disinfection, it is very important to understand that literature reports myriad methods and each one is unique. Therefore, it is difficult to tag any one or few as novel disinfection techniques. Moreover, age-old techniques of disinfection, such as filtration, usage of metals, and so forth, are being tried all over again as such or with some variation. This chapter makes an attempt to discuss some of the methodologies that are unique, different, and have not been described in the chapters before.

6.2 New Generation Techniques for Water Disinfection

6.2.1 Plasma Technologies (Dors, 2011)

Plasma is a fourth state of matter consisting of ionised gas molecules, neutral molecules, radicals, and electrons. All the species in plasma are in the excited state; that is, they have more energy than in the ground state. Plasma processing involves chemical and physical reactions between particles and solid surfaces in contact with the plasma. The outstanding properties of most of the plasmas applied to processing of materials are associated with nonequilibrium conditions. Opportunities for materials processing stem from the ability of a plasma to provide a highly excited medium that has no chemical or physical counterpart in a natural, equilibrium environment. Plasmas alter the normal pathways through which chemical systems evolve from one stable state to another, thus providing the potential to produce materials with properties that are not attainable by any other means. The industries in which plasma processing is vital include semiconductor, aerospace, textiles, metal and ceramic, and waste management in chemical industries, biomedical industries, and many more. The processes involved generally are etching, surface activation for deposition of films and coating surfaces, and cleaning, and for conducting chemical reactions, and so forth.

In respect to the chemical industry, plasma processing is widely used in waste treatment and reforming and processing of fuels. Due to the presence of highly energetic species, plasma can even be used to treat hazardous chemicals that are otherwise very difficult to process. The main advantage of plasma processing is that it allows recovery of materials or energy from wastes in solid, liquid, or gaseous form. It is possible to use plasma processing, due to unique characteristics of plasma, to synthesise industrial chemicals from raw materials never used before.

6.2.1.1 Fundamentals of Plasma

6.2.1.1.1 Classification of Plasma

There are two main types of plasmas, atmospheric pressure plasma and low-pressure plasma. For atmospheric pressure plasmas, the mean free paths between electrons and heavy particles are extremely short, and therefore the plasma is collision dominated. Under such conditions, local thermodynamic equilibrium (LTE) may prevail, which includes kinetic equilibrium ($T_e \approx T_h$, where T_e = electron temperature and T_h = heavy particle or sensible temperature), as well as chemical equilibrium; that is, particle concentrations in LTE plasmas are only a function of temperature. In contrast, in low-pressure plasmas, the mean free paths are much longer, and therefore collisions between particles are much less frequent. Under these conditions, the electron temperature is much higher than the heavy particle temperature, that is, $T_e \gg T_h$ (Figure 6.1).

Even though ionisation in low-pressure plasmas is very high, the gas density in this type of plasma is extremely low. Therefore, thermal equilibrium cannot be achieved between electrons and heavy particles during collisions. Consequently, the heavy gas particles remain cold even after collisions. Plasmas produced in various types of glow discharges, in low-intensity, high-frequency discharges, and in corona discharges are typical examples of cold plasmas.

Within atmospheric pressure plasmas, there are two distinct categories, thermal and nonthermal. In thermal plasmas $T_e \approx T_h$ (LTE exists). The core gas temperatures in thermal plasmas are well above 10,000 K and the gas is

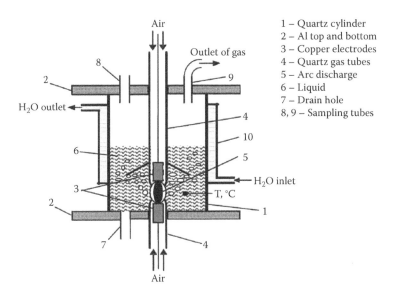

FIGURE 6.1
Plasma processing unit. (From http://www.plastep.eu/fileadmin/dateien/Events/2011/110725_Summer_School/Plasma_water_treatment.pdf.)

significantly ionised. The atmospheric nonthermal plasmas have very high electron temperatures, T_e, while the sensible temperatures, T_h, remain ambient. Atmospheric nonthermal plasmas have a low degree of ionisation and the density of charged species is low. The electrons and ions never achieve local thermodynamic equilibrium. For this reason, the gas is at room temperature. However, atmospheric nonthermal plasmas have a high density of activated species, that is, reactive free radicals and excited state atoms. Thus, nonthermal plasmas are very reactive.

Thermal plasma reactors offer a range of other advantages, including:

1. High throughput with compact reactor geometry.
2. High quench rates (>106 K/s) allowing specific gas and solid material compositions to be obtained.
3. Low gas flow rates (except for nontransferred plasma devices) compared to the combustion of fossil fuels, thereby reducing the requirements for off-gas treatment.

A possible disadvantage, especially from an economic perspective, is the use of electrical power as the energy source. However, a complete comparative cost evaluation often demonstrates the economic viability of plasma-based technologies (Dors, 2011).

6.2.1.1.2 Generation of Plasma

Plasma is generated by the passage of an electric current through a gas. Since gases at ambient temperatures are excellent insulators, a sufficient number of charge carriers have to be generated to make the gas electrically conducting. Passing an electrical current through an ionised gas leads to a phenomenon known as gaseous discharges. Such gaseous discharges are the most common, though not the only, means for producing plasmas.

6.2.1.1.2.1 Thermal Plasma Generation Thermal plasma generation can be achieved using a direct current (DC) or an alternating current (AC) electrical discharge, or using a radio frequency (RF) induction or a microwave (MW) discharge. A DC arc discharge provides a high energy density and high temperature region between two electrodes, and in the presence of a sufficiently high gas flow, the plasma extends beyond one of the electrodes in the form of a plasma jet. The arc plasma generators can be divided into a nontransferred arc torch and a transferred arc torch. In a nontransferred arc torch, the two electrodes do not participate in the processing and have the sole function of plasma generation. In a transferred arc reactor, the substance to be processed is placed in an electrically grounded metallic vessel and acts as the anode; hence, the reacting material should be electrically conductive. Transferred arc torches have been widely used in metallurgical processing. Arc torches and electrodes are usually water cooled, and the average lifetime of the electrodes ranges between 200 and 500 h of operation under oxidative conditions. DC arc plasma torches are commonly available at power levels up to 1.5 MW. Scale-up is possible to 6 MW.

RF plasma torches utilise inductive or capacitive coupling to transfer electromagnetic energy from the RF power source to the plasma working gas. They are very compact and deliver extraordinarily high-input energy per unit volume. Unlike DC arc plasma torches, there are no locally high temperature arcs, no moving parts, and no parts subject to wear. RF current and microwaves can be transferred through insulators, so the use of external electrodes is possible. In this way, the electrodes are not exposed to the severe conditions of thermal plasmas, and therefore have a very long lifetime. RF plasma generators are commonly available at power levels of 100 kW. Scale-up has been demonstrated to the 1 MW range. Many thermal plasma processes have used DC plasma generators due to the stable arcs, but this kind of plasma generator requires expensive electronics and controls, and the plasma plume is very narrow. RF inductively coupled plasma torches are being increasingly considered for a wide range of applications in the area of materials processing. Generally speaking, a RF plasma can generate a very diffuse plume, and the design of external electrodes favours the injection of feedstock material directly into or through the plasma region. However, RF plasma systems often utilise oscillator electronics, which have inherently low efficiencies.

6.2.1.2 Plasma Processing in Waste Treatment

6.2.1.2.1 Thermal Plasma in Waste Treatment (Gomez et al., 2009)

The waste from chemical industries includes volatile organic compounds (VOCs), fly ash and dust particles, polymers, gaseous emissions or liquid effluents, and so on. The treatment methods depend upon the type of wastes; for example, three main options for waste disposal are: (1) burial, (2) treatment followed by burial, and (3) recycling to recover raw material and energy followed by disposal of residue.

With the amount of available land shrinking, burial is becoming a less viable option. Incineration was once a treatment option, but it has technological limitations, for example, treatment of large off-gas volumes and generation of fly ash. The other thermal plasma treatment of wastes have the following advantages:

1. The high energy density and temperatures associated with thermal plasmas, and the correspondingly fast reaction times, offer the potential for a large throughput with a small reactor footprint.

2. The steep thermal gradients in the reactor permit species exiting it to be quenched at very high rates, allowing the attainment of metastable states and nonequilibrium compositions, thereby minimising the reformation of persistent organic pollutants (POPs).

3. Plasmas can be used for the treatment of a wide range of wastes, including liquids, solids, and gases.

4. The high heat flux densities at the reactor boundaries lead to fast attainment of steady-state conditions. This allows rapid start-up and shutdown times, compared with other thermal treatments, such as incineration, without compromising refractory performance.

5. Oxidants are not required to produce the process heat source, as no fuel is combusted; therefore, the gas stream volume produced is much smaller than with conventional combustion processes, and so is easier and less expensive to manage.

6. The combination of the above characteristics allows plasma treatment to be integrated into a process generating hazardous wastes.

Plasma reactors can be employed to melt or, with the addition of glass formers, to vitrify waste to form a stable, nonleachable, glassy slag product in which hazardous substances are trapped within the glass network. The vitrified product offers the potential for reuse, and other products with high added value, such as scrap metals, can be safely recovered. In addition, plasmas can thermally decompose hazardous organic compounds into simpler, benign materials. Alternatively, using gasification or pyrolysis, the organic fraction of waste can be converted into a synthetic gas (syngas) that can substitute for fossil fuels.

The major disadvantage of the plasma process is the use of electricity, which is an expensive energy source, but the use of transferred arc devices means that power is used efficiently and there is no parasitic load associated with the heating of air, with its high nitrogen content, to support combustion, and thus it can be considered a viable long-term solution for waste management.

6.2.1.2.1.1 General Considerations   The high process temperatures mean that volatile metals vapourise and are carried out of the unit, together with halogens and other acid gases, in the off-gas stream, which can be removed using basic slag with a high halogen ion capacity. The material of construction of the air management system unit must be designed to separate, collect, or chemically treat the materials entrained in the off-gas. For waste treatment, hollow graphite electrodes are normally used to produce the plasma arc. The electrodes and the lining of the treatment vessel or chamber are slowly abated or consumed during waste processing, the consumption rate being typically <5 kg/MWh, and therefore an order of magnitude lower than conventional arc furnaces. Plasma treatment units consist of several subsystems besides the thermal plasma source. These components are a waste feed system, a processing chamber, a solid residue removal and handling system, a gas management system, operational controls, and data acquisition and monitoring. Figure 6.2 shows the schematic diagram for plasma waste treatment.

6.2.1.2.1.2 Fly Ash Treatment (Mohai and Szepvolgyi, 2005; Gomez et al., 2009)   Fly ash consists of finely divided particles that are removed before any further treatment of the gaseous effluents. The metal industries generate a considerable amount of fly ash powder. Chemical industries also generate large amounts of waste through processes such as erosion of the catalytic bed, waste incinerators, crushing and grinding operations, and so on. Air pollution control residues are solids obtained from waste gas treatments such as wet acid scrubbing, cyclone separators, and so forth. They contain highly toxic heavy metals such as lead, cadmium, and mercury, and also pose as an environmental threat as they act as catalysts for formation of pollutants in air. Here, recovery of metals from fly ash generated in steel making using radio frequency thermal plasma is discussed as an illustration.

6.2.1.2.1.3 Treatment of Particulate Metallurgical Wastes in Thermal Plasmas   RF plasma systems can process fine powders without granulation in a continuous operation. This possibility, together with the advantageous features of the thermal plasmas mentioned above, offers a great perspective for the synthesis of special ceramic powders such as spinel ferrites. The composition of converter flue dust (CFD) and sludge from the hot plating of steel products (HPS) is given in Table 6.1. Waste samples were treated in reducing atmosphere, respectively, to produce fine metallic powders and in neutral/oxidising atmosphere to synthesise zinc ferrite nanopowders. Table 6.1 shows the bulk chemical composition of CFD and HPS samples.

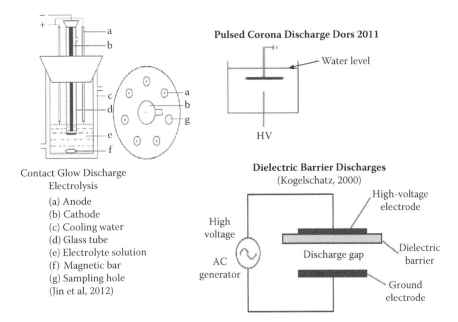

FIGURE 6.2

Types of electrical discharges for water treatment. Contact glow discharge electrolysis: (a) anode, (b) cathode, (c) cooling water, (d) glass tube, (e) electrolyte solution, (f) magnetic bar, and (g) sampling hole (Jin et al., 2012). Pulsed corona discharge (Dors, 2011). Dielectric barrier discharges (Kogelschatz, 2000). (From Jin, X., et al., *Electrochim. Acta* 59(1):474–478, 2012; Dors, M., Plasma for Water Treatment, 2011, http://www.plastep.eu/fileadmin/dateien/events/2011/110725_summer_school/plasma_water_treatment.pdf, accessed May 16, 2012; Kogelschatz, U., Fundamentals and Applications of Dielectric-Barrier Discharges, 2000, http://www.coronalab.net/wxzl/plasma-16.pdf, accessed May 10, 2012).

The experimental results showed that metallurgical waste powders containing iron and zinc oxides and hydroxides can be reduced to metals in RF thermal plasma in the presence of hydrogen. However, if the particles are agglomerated, they cannot be reduced due to the short residence time in the hot plasma region. Thus, to achieve high reduction, nonagglomerated powders with a mean particle size of less than 20 μm should be introduced into the plasma reactor. Spinel ferrites of variable composition were formed with complete conversion during a very short residence time, compared to the

TABLE 6.1

Bulk Chemical Composition of CFD and HPS Samples

Sample	Fe	Zn	Pb	Ca	Mg	Si	Cl
CFD	65	4.22	0.98	1.96	0.15	1.02	0
HPS	17.5	39.5	0.23	2.76	0.29	2.22	6.99

conventional process, which needs heat treatment of 4 to 6 h at temperatures above 1000°C.

6.2.1.2.1.4 *Thermal Plasma Pyrolysis of Organic Waste (Huang and Tang, 2006)*
6.2.1.2.1.4.1 Principle of Thermal Plasma Pyrolysis

Thermal plasma pyrolysis can be described as the process of reacting a carbonaceous solid with limited amounts of oxygen at high temperature to produce gas and solid products. In the highly reactive plasma zone, there is a large fraction of electrons, ions, and excited molecules together with the high energy radiation. When carbonaceous particles are injected into plasma, they are heated very rapidly by the plasma, and the volatile matter is cracked and cracked products are released, giving rise to hydrogen and light hydrocarbons such as methane and acetylene. Approximately four stages can be distinguished in the thermal plasma pyrolysis process:

1. Very fast heating of the particles as a result of their heat exchange with the plasma jet
2. Explosive liberation of volatile matter from the particles
3. Very quick gasification of the homogeneous phase and rapid heat and mass exchange
4. Further gasification of char particles with various gaseous components

Stage 3 could be replaced by quench technology in order to achieve certain technical purposes such as monomer recovery. Addition of water/steam could be effectively used in stage 4 to promote syngas (H_2 and CO) production. High temperature combined with the high heating rate of the plasma results in the destruction of organic waste, giving rise to a gas and a solid residue with varied properties, depending on the feed characteristics and operating conditions.

6.2.1.2.1.4.2 Effect of Operating Parameters

The distribution of gas and solid product from plasma pyrolysis depends on the operating conditions, such as plasma input power. The amount of gas product can reach up to 70 to 80% of the feed, compared to only 10 to 20% for conventional slow pyrolysis and 30 to 50% for rapid pyrolysis. The important operating parameters affecting the product distribution are as follows.

6.2.1.2.1.4.3 Operating Power

As the operating power is increased, the energy density and plasma discharge zone are increased. The longer plasma discharge zone results in longer residence time for particles in the high temperature zone, which in turn increases the probability of primary radical formation and gas yield. However, specific energy consumption sharply increased above a certain temperature, since with the increase of input power, the maximum temperature achieved in the discharge zone increased only slightly because of increased loss of the discharge power as radiation. Also, in the given range of parameters, the gaseous product composition of plasma pyrolysis is mainly controlled by the C/H ratio of the different materials, and hence an excessive increase of input power does not have any effect on product yield. A suitable input power level should be chosen for a particular plasma pyrolysis process in order to obtain reasonable product yield using minimum specific power consumption.

6.2.1.2.1.4.4 Powder Size

Solid conversion and final product formation take place in very short times due to the high enthalpy of the plasma, the rate of heat exchange of the particles with the plasma jet, and the volatile disengagement from the particles (stages 1 and 2), which may themselves exert a certain influence on the conversion. In the plasma reactor, the particle speed depends on both the plasma jet speed and the distance to which the powders are injected into the plasma jet. The particle temperature is a function of the plasma enthalpy and of the residence time in the plasma jet. Besides, the speed and temperature of the particles are influenced by their dimensions, specific heat capacity, and density. A small particle size and a high heating rate are preferred for a higher conversion to gaseous product. These particles could be suspended in gases or liquids. These interactions with the plasma jet are similar, except the power requirement and the final product formation are different. Table 6.2 shows the product distribution with and without the quench process.

Thermal plasma pyrolysis can also be used as a depolymerisation process and subsequently monomer recovery from polymer waste when combined with a quench process. In plasma pyrolysis stage 3, complex homogeneous reactions take place after depolymerisation. In order to preserve monomer molecules in the low-temperature gas, a quenching process is needed to avoid further dissociation of the monomer molecules. Table 6.3 shows the

TABLE 6.2

Product Distribution with and without Quench Process

Product	With Quenching Feed: PE	With Quenching Feed: PP	Without Quenching Feed: PP
Hydrogen	N.A	N.A	62
Propylene	53.39	93.7	N.A
Ethylene	22.24	1.7	N.A
Methane	5.62	2.6	8.0
Acetylene	5.7	N.A	21.4
1,3-Butadiene	3.65	N.A	N.A
C_nH_m and unknown	N.A	N.A	6.8
Butanes and butenes	N.A	1.3	N.A
Solid conversion	50	78	94

TABLE 6.3

Product Distribution from Thermal Plasma Pyrolysis of PP with and without Steam Injection

Product	Concentration (vol%) without Steam Injection	Concentration (vol%) with Steam Injection
H_2	14.99	27.06
CO	0.83	13.33
C_2H_2	4.96	3.87
CH_4	1.58	1.45
C_2H_4	1.07	0.57
C_2H_6	0.07	0.02
N_2	71.83	49.02
C_nH_m + unknown	4.17	4.37
Solid conversions	94%	96%
Gas yield	885 ml/g	2167 ml/g

comparison between plasmas with quenching and without quenching on the basis of conversion of feed to gas and product distribution in an inductively coupled plasma (ICP) reactor. The gas distribution is given in terms of percent volume on carrier-free basis (N_2 or Ar).

In ICP pyrolysis of polyethylene (PE), the amount of propylene was much higher than expected and was believed to be due to the β-scission reaction mechanism occurring at the high temperatures. Hence, additional experiments involving variations in the residence time of the polymeric powder in the plasma zone should be conducted to obtain higher yields of monomer in the product gas stream. In addition, more work is required to investigate various types of quenching systems.

6.2.1.2.1.4.5 Steam Reforming

Experimental data listed in Table 6.3 show that using steam reforming, synthesis gas can be produced from waste polymer by thermal plasma pyrolysis.

An essentially large volume of CO is generated under the water/steam plasma compared to under nitrogen or argon plasma, and at the same time, H_2 concentration and total gas yield are increased significantly. Beside reactions of the gaseous hydrocarbon and steam, the reaction between carbon and steam is believed to play an important role in the process according to the reaction $C + H_2O \rightarrow CO + H_2$.

6.2.1.2.1.5 Product Characteristics
6.2.1.2.1.5.1 Gas Distribution

Table 6.3 shows the gas distribution for various feeds. Large quantities of combustible gases, such as H_2, CO, C_2H_2, CH_4, and C_2H_4, were produced during the pyrolysis processes. The concentrations of each component can vary considerably according to the feed characteristics and pyrolysis process parameters. Pollutant gases, such as SO_2 and NO_x, are only found in low concentrations. The combustion heat value of the gas product is in the range of 4 to 9 MJ/Nm^3, so it can be used directly as a fuel in various energy applications, such as direct firing in boilers, gas turbines, or gas engines.

6.2.1.2.1.5.2 Char

Depending on the plasma pyrolysis conditions, the char fraction contains inorganic materials ashed to varying compounds, any unconverted solid additives such as carbon black filler in tires, and carbonaceous residues formed from thermal decomposition of the organic components, and even carbon nanoparticles produced in the gas-phase secondary reactions.

Char can be used as solid fuel due to its heating value, but for better process economy, other high-value applications of solid products are sought, depending on their characteristics. Guddeti, Knight, and Grossmann (2000) reported that the solid residue from pyrolysis of plastic contained almost 100% carbon with several novel structures, indicating the potential of several high-value applications of this solid carbon, such as production of high surface area catalysts, carbon adsorbent, and so forth. In the treatment of waste tires using thermal plasma pyrolysis, one attractive option is to recycle the maximum amount of carbon black filler that can be used as low-grade semireinforcing filler for nontire applications and cannot be used directly as a carbon black source for tire manufacture due to its high ash content. High-grade carbon black may be obtained after further processing of the solid residue mixture by acid and base wash, and de-ashing.

6.2.1.2.1.6 Process Advantages and Application Potential Thermal plasma pyrolysis treatment of organic waste has the following main advantages:

1. Efficient delivery of heat energy for simultaneous rapid promotion of both physical and chemical changes in waste material.
2. The properties of plasma pyrolysis products are suitable for energy and material recycling. Thermal plasma pyrolysis of organic waste gives only two product streams—a combustible gas and a solid residue—both of which are useful, easy-to-handle products, compared to conventional pyrolysis, which leads to gas, liquid, and solid products. The liquid product is a tarry oil consisting of a variety of heavy hydrocarbon compounds, and separation and collection of the oil from other gas and solid products is difficult.

A preliminary analysis indicated that plasma pyrolysis treatment of rubber waste has economic potential, given the following conditions are met:

1. Capital investment for a plant processing rubber waste at 300 kg/h is $1,500,000.
2. Specific energy consumption is 1 kWh/kg rubber feed.

3. The electricity price for the industrial sector is $0.05/kWh.

4. Carbon black recovery is 23 wt% of rubber feed; the market price for semireinforcing carbon black is $500/ton.

5. The gate fee for receiving rubber waste is $30/ton.

6. The gas yield is 3 nm³/kg rubber feed, and with a calorific value of 9 MJ/nm³, the gas is combusted in a boiler or gas engine for power generation at a minimum efficiency of 26%.

6.2.1.2.2 *Nonthermal Plasma in Waste Treatment (Subrahmanyama et al., 2010)*

6.2.1.2.2.1 *Volatile Organic Compound (VOC) Abatement: Catalytic Nonthermal Plasma Reactor for Abatement of Toluene*
For the abatement of dilute VOCs (<1000 ppm), conventional techniques like adsorption, thermal, and thermocatalytic oxidation are not suitable, mainly due to high energy consumption. Among the alternatives, nonthermal plasma (NTP) generated at atmospheric pressure appears to be the energy-saving approach. Nonthermal plasma chemical processing of hazardous air pollutants (HAPs) has been extensively investigated, and it has been demonstrated as an efficient technique for the decomposing of olefinic HAPs. NTP is a combination of energetic electrons, radicals, ions, and excited species as well as radiation. In NTP, the electrical energy is primarily used for the production of energetic electrons without heating the flue gas.

Here, NTP generated by using a dielectric barrier discharge (DBD) using either a conventional metallic rod or transitional metal oxide–modified sintered metal fibre (SMF) as an inner electrode for abatement of toluene is discussed. The SMF was modified with 3 wt% of Mn and Co oxides via impregnation of aqueous metal nitrate solutions followed by calcination at 773 K for 4 h. In DBD, the presence of dielectric distributes the microdischarges throughout the discharge volume. These microdischarges initiate the chemical reactions in the gas phase through electron impact dissociation and ionisation of the carrier gas. However, NTP abatement of VOCs shows low selectivity (0.3 to 0.5) to total oxidation (CO_2 and H_2O) and may result in the formation of undesired and sometimes toxic by-products. In order to improve the efficiency of NTP, often a catalyst is combined with the plasma technique.

6.2.1.2.2.2 *Performance of a DBD Reactor*
During the oxidative decomposition of toluene, the desired products are CO_2 and H_2O. However, in general NTP may lead to undesired products, and hence the selectivity to CO_2 is not 100%. The performance of the reactor is therefore evaluated on toluene conversion and selectivity to CO_2, which is calculated as follows:

$$S_{CO}(\%) = \frac{[CO]}{7.([VOC]0-[VOC])} \quad \text{and} \quad S_{CO_2}(\%) = \frac{[CO_2]}{7.([VOC]_0-[VOC])}$$

$$S_{COx} = S_{CO} +$$

The performance depends upon the concentration of toulene in the feed and specific energy input, which is defined as $(J/I) = \frac{\text{power(w)}}{\text{gasflowrate}}$.

It was observed that for any feed concentration, as SIE increases the conversion and selectivity, both increase for every type of inner electrode.

It can also be observed that as the concentration of feed decreases, both selectivity and conversion increase, and also at 1000 ppm there is no improvement in performance by SMF modification with transition metal oxides. This could be due to quick deactivation of the catalyst by polymeric products formed in the reaction at high feed concentration. For 100 ppm feed with MnOx/SMF catalyst, at 17.5 kV, the conversion was ~100% and there was no polymeric carbon deposit. The NTP reactor with catalytic electrodes shows better performance than the plasma reactor using conventional Cu or SMF electrode. Modification of SMF by MnOx and CoOx increased the product selectivity toward total oxidation (CO_2 and H_2O). The better performance of CoOx and MnOx/SMF may be due to the formation of atomic oxygen by in situ decomposition of ozone on the surface of the catalytic electrode.

6.2.1.2.2.3 Advantage of Catalysts Advantages of using plasma-catalysis systems over NTP include an increase in the conversion of VOC at low input energy, a higher selectivity to total oxidation, and a decrease in undesired by-products. VOCs can be decomposed at ambient temperatures, where in general the catalyst will not be active.

6.2.1.2.2.4 Configuration of Catalytic Plasma Reactor The catalytic plasma reactor can be configured in the following two ways:

1. In plasma reactor (one stage): The catalyst is placed in the discharge zone.
2. Post plasma reactor (two stages): The catalyst is placed downstream of the discharge zone.

The former shows better performance, probably due to a better reaction condition, like ready availability of other short-lived oxidising species and ozone. In post plasma configuration performance, an increase is only due to the oxidising properties of long-lived species, mainly ozone and oxides of nitrogen. Short-lived oxidising species such as oxygen radical anion O_n^{-1} and activated O_2^* cannot reach the catalyst surface in a downstream configuration. In one-stage configuration, the synergistic effect may be due to an increase in the concentration of short-lived excited species on the surface of the catalyst and the effect of photons and electrons generated in the plasma. In addition, plasma catalytic systems have been shown to follow zero-order kinetics during VOC decomposition, indicating the importance of surface reactions; hence this configuration is expected to show promising results.

6.2.1.2.2.5 Removal of Other Compounds Using Nonthermal Plasma Apart from VOCs, nonthermal plasma is used to remove harmful air pollutants like NO_x, SO_x, and so forth. It is also used for treatment of toxic waters from textile industries and to treat gases like SF_6, perfluoro carbons and chlorofluoro carbons. A pilot plant study showed that the corona plasma-induced oxidation technique can be applied industrially to convert concentrated ammonium sulphites to sulphates in an aqueous solution and efficiency of 20 to 60% after one cycle was reported when the sulphite concentration was around 1–3 mol/l. Multiple cycle treatments can be used to improve the oxidation efficiency. The energy consumption is around 20 $Whmol^{-1}$.

6.2.1.3 Treatment of Polluted Waters

There is a continuing need for the development of effective, cheap, and environmentally friendly processes for the disinfection and degradation of organic pollutants from water. Ozonation processes are now replacing conventional chlorination processes because ozone is a stronger oxidising agent and a more effective disinfectant without any side effects (Malik et al., 2001). However, the fact that the cost of ozonation processes is usually higher than chlorination processes is their main disadvantage. Hence, water treatment by direct electrical discharges may provide a means to utilise these species in addition to ozone. The units based on electrical discharges in water or close to the water level are being tested at industrial-scale water treatment plants (Malik et al., 2001). Mainly three methods, pulsed corona discharge, DBD, and contact glow discharge electrolysis techniques, are being studied for the purpose of cleaning water. In the area of water purification, ozone synthesis is an industrially accepted application of electrical discharges. Ozone is required in huge quantities for drinking water and wastewater treatment. The major advantages of the ozonation process over conventional chlorination processes for water treatment have been discussed in Chapter 2 on chemical disinfection techniques. There is no need to store and handle toxic chemicals; it is a stronger and faster-acting oxidiser and can safely destroy a broader range of organic contaminants. It also helps in removal of colour, odour, and suspended solid materials, and ozone is far more efficient in killing bacteria, viruses, spores, and cysts—some of the merits of using ozone for water disinfection. If the reactor is designed to create electric discharges just below or at the water surface, some of the reactive radicals can get inside water and bring about disinfection. This also implies that a separate electrical discharge reactor to generate ozone by the corona discharge method (already described in Chapter 2) and tubings to carry ozone-rich air will not be required. Therefore, instead of ex situ electrical discharges for ozone production, the in situ electrical discharges in water may provide a means to utilise most of these chemically active species for water cleaning. Furthermore, the intense electric fields necessary for electrical discharges are also lethal to several kinds of microorganisms found in water and show a synergistic lethal effect when combined with conventional disinfectants such as O_3 and H_2O_2. The electrical discharges in water may also produce UV radiation and shock waves, which help in the destruction of pollutants. For these reasons direct electrical discharges in water are clearly the best next-generation technologies for water treatment.

They are environmentally friendly and may prove to be far more effective than conventional oxidants and disinfectants. Techniques of direct electrical discharges in water and the electrical discharges in close proximity to the water surface are being rapidly developed and tested on the industrial scale for water and wastewater treatments. On the other hand, direct electrical discharges in water need much higher electric fields and more complicated power sources.

Three types of electrical discharges are usually reported for water purification (Malik et al., 2001):

1. Contact glow discharge electrolysis
2. Dielectric barrier discharges (also called silent discharges)
3. Pulsed corona discharges

Figure 6.2 describes the types of electrical discharges for water treatment.

In contact glow discharge electrolysis a continuous DC voltage of around 0.5 kV is applied to a thin wire anode in contact with the water surface, while the cathode is dipped in water and isolated from the anode through porous glass. A sheath of vapour forms around the anode through which current flows as a glow discharge. Charged species in the plasma (present in the discharge gap or sheath of vapour around the anode) are accelerated due to the steep potential gradient and enter the liquid phase with an energy that may be as high as 100 eV. In the case of contact glow discharges, almost all the species in the discharge zone heat up, so that the plasma generated in the reactors can be called hot plasma. In silent discharges and pulsed corona discharges, like described below, only free electrons gain high energy, and the rest of the heavier charges and neutrals remain close to room temperature; the plasma generated in this way is called cold plasma or nonequilibrium plasma.

In a *dielectric barrier discharge* reactor the electrical discharges take place between electrodes where at least one of the electrodes is covered with a thin layer of dielectric material, such as glass or quartz. In the case of the water treatment application of dielectric barrier discharge reactors, a layer of water around one of the electrodes acts as a dielectric. Usually an AC voltage of around 15 kV is applied across the electrodes.

In both cases of contact glow discharge electrolysis and dielectric barrier discharge reactors the electrical discharges take place in the gas phase in close proximity to the water surface. They require an intense electric field for electrical discharge to take place in water. Such a high electric field is possible by applying high-voltage pulses of 15 to 100 kV, usually of positive polarity, with a sharp rise time (a few nanoseconds) and short duration (nano- to microseconds) in a pulsed corona discharge reactor. Furthermore, the pulsed corona discharges are effective disinfectants, and they can also take place in the gas phase in close proximity to the water surface. This is why most of the studies on water treatment are carried out using pulsed corona discharge reactors and the available industrial-scale units are also based on this technique.

A *pulsed corona discharge reactor* requires a pulse generator and a reactor. The pulse generator is commonly based on the discharge of a capacitor on a low-inductance circuit through a spark gap switch. The reactor is comprised of metallic electrodes and fittings made of some insulating materials. The electrodes are usually in a needle-plate arrangement, where a needle is connected to the high-voltage terminal and the plate is earthed. The needle is covered with an insulator, for example, Teflon, and only its tip is exposed, so that an intense electric field may develop at the needle tip.

6.2.1.3.1 Drinking Water Treatment by Electrohydraulic Discharge

Electrohydraulic discharge (ED) is a direct plasma technology that has the ability to treat a wide range of aqueous contaminants within a single unit process since it combines both radical and UV processes. The process involves pulsed arc discharges within the water to be treated. It leads to a variety of physical and chemical processes, such as UV irradiation, radical reactions, electron processes, ionic reactions, thermal dissociation, and pressure waves, to name a few (Chang, 2001; Locke et al., 2005). High-voltage electrical discharges directly in water (electrohydraulic discharge) or in the gas phase above the water (nonthermal plasma) have been demonstrated to produce hydrogen peroxide, molecular oxygen and hydrogen, hydroxyl, hydroperoxyl, hydrogen, oxygen, other radicals, and with the addition of air or oxygen at the high-voltage electrode, ozone. In addition, depending upon the solution conductivity and the magnitude of the discharge energy, shock waves and UV light may also be formed. In turn, these reactive species and physical conditions have been shown to rapidly and efficiently degrade many organic compounds, such as phenols, trichloroethylene, polychlorinated biphenyl, organic dyes like methylene blue, aniline, anthraquinone, monochlorophenols, methyl *tert*-butyl ether (MTBE), benzene, toluene, ethyl benzene (BTEX), and 2,4,6-trinitrotoluene, 4-chlorophenol, and 3,4-dichloroaniline. In addition, the oxidation of several inorganic ions in water has been studied with various electrical discharge processes.

There are two hypotheses believed to be the mechanism of action of the plasma technology. The first one favours an electron multiplication theory in the liquid. In the past, it was often believed that a current multiplication mechanism such as the development of electron avalanches in gas discharges to initiate breakdown. Even more direct correlation between these avalanches and the consequent formation of vapour bubbles in the liquid has been demonstrated. However, electron avalanche processes in bulk water are negligible due to the usual small high electrical field region near the metal electrode and the large scattering cross sections that make it almost impossible for the electrons to gain sufficient kinetic energy for impact ionisation. Additionally, free electrons are generally absent in water because even if they are present, they are quickly solvated. Hence, the probability of free electrons in the bulk water is negligible, although one must be careful not to generalise ideas for different liquids and not to exclude electron avalanche processes without a good motivation. The second one favours *bubble mechanism breakdown theory* or, more generally, a phase change mechanism breakdown theory. A general acceptance is growing that preexisting bubbles and field enhancement effects in the near-electrode region are involved even for nanosecond voltage pulse widths. Bubbles can preexist due to dissolved gases or can be generated by local heating and resulting degassing. The potential advantages of the electrical discharge process are manifold (Locke et al., 2005).

Electrical breakdown is generally defined as the moment when a conductive plasma channel forms an electrical connection between the two metal electrodes inside the liquid. This leads to the formation of a spark or arc. A time lag between application of the high voltage and breakdown is always observed. This time lag consists of three successive steps: initiation phase or streamer inception, streamer propagation phase, and spark and arc phase. On a small scale, the treatment of water is conducted in a glass tube reactor (inner diameter of 22.5 mm) equipped with water pumping and cooling systems. A pulsed positive spark discharge is generated between a high-voltage stainless steel hollow needle electrode and a grounded rod electrode, both immersed in the water separated by a gap of 7 mm. The role of the spark discharge plasma is the generation of highly oxidative hydroxyl radicals, UV radiation, and shock waves. All these factors have a biocidal effect.

6.2.1.3.1.1 Biocidal Effect of Electrohydraulic Discharges In bacteria, cell destruction of membranes through alteration of glycoproteins or glycolipids (Scott and Lesher, 1963) and certain amino acids such as tryptophan (Goldstein and McDonagh, 1975). Disruption of enzymatic activity of bacteria by acting on the sulphhydryl groups of certain enzymes and their effect on both purines and pyrimidines in nucleic acids (Scott and Lesher, 1963) can also result in cell death. In the case of viruses, modification of the viral capsid sites that the virion uses to fix on the cell surfaces occurs. High concentrations of ozone dissociate the capsid completely (Cronholm et al., 1976; Riesser et al., 1976). Modifications in the oocyst structure are the main mechanism in the case of protozoans.

6.2.1.3.1.2 Disinfection By-Products of Electrohydraulic Discharges   Like other water disinfection techniques, this process also results in the generation of disinfection by-products. Some of them are formaldehyde, acetaldehyde, glyoxal, methyl glyoxal, aldehydes, acids, oxalic acid, succinic acid, formic acid, acetic acid, aldo- and ketoacid, and pyruvic acid.

In the case of water treatment, the main and most significant impacts to the environment are associated with the usage of electrical power. The consumption of electrical power was the highest for the plasma-based electrohydraulic discharge system (25 kWh/m^3 compared to 0.2 kWh/m^3 of UV and PEM ozonation method). In general, plasma-based technologies can be considered equally competitive ones compared to the other researched methods. The relatively high demand of electrical energy causes lower positioning of plasma technologies in cases where no other materials are utilised and major waste is formed. On the other hand, many traditional technologies are associated with high amounts of process waste, which provides plasma technologies with an opportunity to establish them in the market as more efficient and, on many occasions, more environmentally friendly.

6.2.2 Super Sand for Water Disinfection

An interesting breakthrough for cleaning drinking water is the use of "super sand." The technique essentially involves coating grains of sand in an oxide of graphite that is commonly used as lead in pencils. Graphite oxide is essentially a compound of carbon, oxygen, and hydrogen in variable ratios, obtained by treating graphite with strong oxidisers. The maximally oxidised bulk product is a yellow solid with a C:O ratio between 2.1 and 2.9 that retains the layer structure of graphite, but with a much larger and irregular spacing (Lerf et al., 1998). It is indeed a low-cost way of purifying water, especially in the developing world. This finding has been published in the *American Chemical Society Journal of Applied Materials and Interfaces*. According to the World Health Organisation (WHO), 60% of the population in sub-Saharan Africa and 50% of the population in Oceania (islands in the tropical Pacific Ocean) use improper sources of drinking water. Therefore, there is a dire need to address this problem. Sand has been used since ancient times for filtration of drinking water. The technique consists of first dispensing graphite into water and mixing with regular sand. This mixture is heated up to a temperature of 105°C for a couple of hours to evaporate the water. The final product is the coated sand that is used to purify polluted water. It is interesting to note that the graphite oxide can be modified to make it more selective and sensitive to the adsorption of removal of certain pollutants, such as organic contaminants or specific metals in drinking water (Gaot et al., 2011).

6.2.3 Energy-Efficient Pen-Size Disinfection Devices (Khaydarov, Khaydarov, and Yuldashev, 2007)

A remarkable oligodynamic disinfection method employing different metal ions such as Ag^+, Cu^{2+}, and Au generated by an electrolytic process has been researched for water disinfection. The amount of ions used was within the limits laid down by drinking water regulations. Effectiveness for killing *Escherichia coli*, *Legionella pneumophila*, *Salmonella*, *Vibrio cholerae*, and other pathogens was examined.

In the study undertaken, three types of water disinfection devices with the alloy composition of electrodes Ag/Cu/Au in the ratio 79.9%:20%:0.1% were used in the form of a pen-size portable device using a 6 to 12 V battery, devices with a productivity of 1 and 50 m^3/h using a solar battery, and 110 to 220 V AC, respectively. The portable devices were tested over a period of 7 years in different labs and universities, such as in the Central Sanitary and Epidemiological Laboratory of the Ministry of Defence of Uzbekistan, and the Department of Pathology of the Sains University of Malaysia. The disinfecting devices with productivity of 1 m^3/h using solar battery were tested in disinfection of artesian well water containing 102 to 104 CFU/l *E. coli* by JDA International (Grand Junction, Colorado), by the Central Sanitary and Epidemiological Laboratory of the Ministry of Defence of Uzbekistan, and by the Ministry of Public Health of Uzbekistan. Energy consumption of the devices was 0.2 W-h, and 250 devices were installed in manual artesian well water pumps in the Aral Sea region of Uzbekistan for disinfection of underground water. These devices have been used in the United States at the Denver (Colorado) Zoological Park, in Malaysia after the tsunami for water disinfection of two fish farms and one water storage system, and in Russia for treatment of swimming pools.

These studies investigated the dependence of disinfecting time against effective concentration C, initial bacteria concentration N_0 (from 103 to 1012 CFU/l), and organism type. The results of this study indicated that the pen-size devices with capability of 3 m^3/h of treated water can be used by the population to prevent bacterial contamination in emergency situations. The devices, with productivity of 5 to 50 m^3/h, have small sizes, very low energy consumption of 0.5 to 5 W-h, work housing accumulators, solar batteries, or 110 to 220 V AC supply. They can be used for water disinfection of potable water systems, water storage systems, cooling water in air conditioners, swimming pools, and so forth.

6.2.4 Online Monitoring Technologies for Drinking Water Systems

Demand for more efficient and comprehensive online monitoring technologies for water quality assessment is the need of the hour. This may be attributed to the stringent constraints placed on water companies to provide high-quality drinking water and increasing scarcity of water resources. Traditionally, assessment of water disinfection was carried out by sampling small amounts followed by its microbiological analysis. This approach can only capture a small amount of data and may not necessarily detect potentially important events. Therefore, state-of-the-art technologies for online monitoring of water quality may be a potential solution to this issue.

Online monitoring is usually defined as the sampling, analysis, and reporting of a parameter that produces a sequence of data at a much greater frequency than that permitted by manual sampling. It also allows real-time feedback for process control, water quality characterisation, and operational or regulatory purposes. Online monitoring can be carried out on-site as well as at remote locations and will deliver measurements at intervals of seconds to minutes apart. Clearly, online instrumentation must be placed at representative locations in the water system and must be periodically maintained by qualified technical personnel.

Although there are different online monitoring systems for physical, chemical, and hydraulic parameters, the biological parameters reflect the assessment of microbial quality of water. There are two basic types of biological monitors currently in use: those that use biological species as indicators of the presence of contaminants of concern, such as toxic chemicals, and those that screen for the presence of biological species of concern, like algae, pathogens, and so forth. Currently, many existing biological monitors are quite new and can be considered experimental/unique applications. Sensitivity of test organisms to individual compounds must be determined initially. Online biological monitors are a very active area of R&D due to increasing regulatory and public demand pressures. While bacterial-based systems show great sensitivity and ease of operation, development in this area most likely will derive from improved fingerprinting of organisms and maintenance cost reduction. Most advances can be expected from protozoan monitor technology, with techniques in UV absorption/scattering analysis that may soon allow automated detection of *Cryptosporidium* and *Giardia*. Also, molecular techniques initially applied to the recognition of the genomic sequence of specific organisms in clinical applications (Bej, 2003) have also shown great potential for use in the detection of pathogens in water, and are producing extremely interesting results that could lead to widespread online use in the very near future.

6.2.5 Electrocoagulation-Electroflotation

This is a very old technique of water treatment that has been used for over a century. Several large-scale water treatment units employing electrocoagulation were in use in London toward the late nineteenth century (Matteson et al., 1995). The technique basically involves the use of consumable or sacrificial electrodes that erode in water to release ions. These are generally iron or aluminium ions released from the anode. Once they are released, they get hydrolysed into polymeric iron or aluminium hydroxides that are known coagulating agents. They react with negatively charged particles that are brought toward the anode due to the electrophoretic action. Thus, when contaminated water is being treated, the negatively charged pollutants, including the microbial cells, react with the polymeric hydroxides by various chemical and physical processes, leading to their coagulation. They are then removed by electroflotation (explained below) or by conventional sedimentation. The main difference between normal coagulation processes and electrocoagulation is the need to add coagulating agents such as aluminium hydroxide into the water to be treated in the former, whereas in the latter the coagulating agent is generated *in situ* by the electrocoagulation process.

Electroflotation is generally used subsequent to elecrocoagulation during water treatment. It involves the use of electrolytically generated hydrogen and oxygen bubbles for the separation of the suspended solids from a liquid. It has also been employed in effectively treating wastewater, groundwater, industrial sewage, and effluents containing heavy metals on account of better efficiency, easy operation, and less maintenance compared to conventional techniques of dissolved air flotation.

Electrocoagulation-electroflotation has been employed for the treatment of water, and current reports in literature point to its use in the removal of chemical and biological contaminants. In an interesting study, this combination has been used for fluoride removal in drinking water. Fluoride could be effectively reduced from an initial value of 4 to 6 mg/l to 1 mg/l within a span of 30 min. The pH of the water appeared to be very critical for fluoride removal, the optimum being 6 to 7. Moreover, the presence of calcium ions had a positive effect, whereas sulphate had a negative effect on fluoride removal (Zuo et al., 2008). Yet another interesting report on surface water treatment clearly demonstrates the suitability of electrocoagulation-electroflotation for industrial uses. An aluminium soluble electrode was used to treat different water samples from a river and pond. It was observed that the water quality improved markedly. The mechanism of disinfection has been attributed to a combination of two effects. The process of electrocoagulation-electroflotation led to a decrease in molecular oxygen, phosphate, and nitrate anions, as well as dissolved organic compounds, due to which the total microbial flora decreased. The second effect was due to the formation of nanocrystallites that could also have exhibited an antimicrobial action (Ricordel et al., 2010).

6.2.6 Nanotechnology

This topic has already been introduced in Chapter 2, where recent use of zero-valent iron for water disinfection was described. A literature survey of current trends appears to indicate nanotechnology as one of the upcoming water disinfection methodologies in the future, and therefore it is important that it is further emphasised and elaborated here. The highlight here would be on the fundamentals and also some of the recent research in the area, with greater emphasis on the innovative way in which nanotechnology is being researched and implemented for water treatment.

It is very apparent that nanotechnology is expected to provide extraordinary opportunities to develop cost-effective and environmentally acceptable water treatment processes, and several chemical, physical, and biological pollutants can be expected to be removed by the application of nanotechnology. Generally, four classes of nanomaterials are employed for water treatment: dendrimers, zeolites, carbonaceous nanomaterials, and metal containing nanoparticles. All four classes have been found to have a substantially broad property of dealing with the contaminants commonly found in water.

Dendrimers are highly branched macromolecules with a controlled composition. They are basically made of a central core and repeating units with terminal functional groups. The nature of these units and the functional groups determine the dendrimer's properties and its behaviour in the external medium. Dendrimers have been used to enhance the performance of membrane systems used for water treatment. These membranes are reported to have pore sizes in the range of 0.1 to 1 nm. They are very effective in the removal of many organic and inorganic contaminants, including the hardness found in water due to multivalent cations (Zeman and Zydney, 1996). Dendritic polymers exhibit many features that make them extremely suitable as functional materials for water purification. They are available in size ranges from 1 to 20 nm and can be employed as recyclable water-soluble ligands for the removal of toxic compounds like certain metal ions, radionuclides, and inorganic anions (Ottaviani et al., 2000). It is interesting to note that a dendrimer-enhanced ultrafiltration process (DEUF) has been used for recovering metal ions from aqueous solutions (Diallo et al., 2005). In this study, poly(amidoamine) dendrimers with ethylene diamine core and terminal amino groups were tested to recover copper ($Cu(II)$) ions from aqueous solutions. The technique consisted of mixing the dendrimer with the contaminated water, which results in the formation of a metal complex involving the dendrimer and the copper ions in solution. These metal complexes were separated by using ultrafiltration membranes with appropriate molecular weight cutoffs. Regeneration of the complex could be achieved by decreasing the pH to 4, which allowed the metal ions to be separated from the dendrimer. The dendrimer was recycled for use once again. The DEUF systems had the advantage of using smaller operating pressure and hence lower energy consumption (Diallo, 2004).

Zeolites are aluminosilicate minerals that are microporous and find common application as adsorbents, especially in water purification. They also serve as ion exchange media for metal ion removal from acid mine wastewaters (Moreno, Querol, and Ayora, 2001). Similarly, they have been reported to remove other heavy metals, like chromium, nickel, zinc, and cadmium, from metal electroplating effluents (Alvarez-Ayuso et al., 2003).

Carbonaceous nanoparticles can also be used as an effective remedy for polluted waters. They can be employed in various forms, such as nanofibres and nanotubes, and can also be encapsulated in vescicles. In an interesting study, nanoporous activated carbon fibres (ACFs) were used for sorption of benzene and toluene. The average pore size of the fibre was 1.16 nm, and surface area ranged from 171 to 483 m^2/g. It is reported that the sorption followed the Freundlich adsorption isotherm and the equilibrium constants obtained were much higher than those of granular activated carbon (Mangun et al., 2001).

Carbon nanotubes (CNTs) are allotropes of carbon with a cylindrical nanostructure. They have unusual properties that are valuable for water purification. They may be single-walled nanotubes (SWNTs) and multiwalled nanotubes (MWNTs). Dichlorobenzene adsorption using CNTs has been studied. It was observed that 40 min is required for maximum sorption of 38.3 mg/g (Peng et al., 2003). It has been reported that multiwalled carbon nanotubes are better sorbents of volatile organic compounds than carbon black (Li, Yuan, and Lin, 2004). Cross-linked alginate vescicles have been used for encapsulation of these multiwalled carbon nanotubes, and these caged nanotubes showed better sorption capacity in a study conducted on four water-soluble dyes (Fugetsu et al., 2004).

The role of metallic compounds as disinfectants has been dealt with in Chapter 2. Since silver has antimicrobial properties, researchers started evaluating silver nanoparticles as biocides. They have been found to be effective against a wide range of gram-positive and gram-negative organisms such as *Staphylococcus aureus*, *E. coli*, *Klebsiella pneumonia*, and *Pseudomonas aeruginosa* (Son, Youk, and Park, 2006). Use of other metals, especially zero-valent iron nanoparticles, has been discussed in previous chapters.

6.2.7 Solar Disinfection (SODIS)

Solar disinfection is yet another water disinfection technique that is being discussed again on account of its current relevance as well as for the future. This topic has been explained in detail in Chapter 3, where its fundamentals, mechanism of disinfection, factors affecting the solar disinfection, and several current modifications and additions to SODIS from literature have been cited. In this section the latest developments are discussed.

In one of the recent studies, the effect of additives as well as alternative container materials for SODIS has been explored. The study was carried out on both viruses and bacteria. The organisms investigated were MS2 coliphage,

E. coli, and the *Enterococcus* species. The combinations of additives studied, that is, sodium percarbonate in combination with either citric acid or copper plus ascorbate, successfully improved SODIS. Various container materials, such as bottles made of polypropelene copolymer (PPCO), resulted in 3-log inactivation of these organisms in around half the time required for the same level of inactivation when bottles of polyethylene terephathalat (PET) were used. It is very interesting to note that the microbes from wastewater took a longer time to be killed than laboratory-grown culture, which points out the fact that naturally occurring organisms may be a challenge to tackle and lab results can never be taken for granted (Fisher et al., 2012). Natural calamities and disaster can give rise to situations where safe drinking water may be difficult to access. In such cases it is a challenging task to decontaminate water due to lack of facilities and infrastructure. On-site treatments are favourable in such situations, and SODIS can be a good choice for water decontamination (Loo et al., 2012). Another important area that is being researched in the field of SODIS is the suitability of PET bottles and the toxicity associated with plastic containers. It is known that regulations govern the use of only certain materials in the containers that can be in contact with the foodstuff stored in them. The same applies for drinking water. In a recent study, the effects of temperature, UV exposure, and frequency of PET bottle reusage on the extent of leaching of antimony and bromine were investigated. It is concerning to note that the frequency of reusing PET bottles determined the amount of antimony and bromine leaching into the water. The highest amount reported is approximately 360 ng/l when the bottle was reused 27 times. However, this concentration of antimony in water is well within the acceptable limits and did not pose any risk to the consumer. Similarly, the amount of bromine that leached into water (approximately 15 μg/l) did not pose any risk to human health. However, the investigators state that there are no acceptable limits that are laid down for organobromine compounds. Therefore, if these are detected in water, it would be very difficult to predict the risk involved (Andra et al., 2011).

These and several studies related to leaching of potentially harmful compounds from PET bottles used for SODIS have caused concerns. However, there has been some controversy regarding the presence of some genotoxic and estrogenic substances that have been derived from PET bottled water. Formaldehyde, acetaldehyde, and antimony are clearly related to migration from PET into water, but their origin has not been clearly understood. It appears that contradictory results are reported in the literature owing to the variety of analytical techniques, bioassays, and different exposure conditions used (Bach et al., 2012).

Yet another relevant aspect of current SODIS research is directed toward large-scale and continuous water treatment, especially in developing countries. In one such investigation, a continuous-flow solar UVB disinfection system was developed and its effectiveness was tested to inactivate *E. coli*, the indicator of pollution in drinking water. Mathematical simulations were used to indicate whether incorporation of a device to amplify the UVB fluence rate would benefit the flow system. Indeed, results pointed out the positive influence of the same, and a compound parabolic collector was used for the same, the material of construction being easily available and cheap in many developing countries. SODIS was accomplished by passing water through a UVB transparent pipe that was positioned in the focal area of the compound parabolic collector. This allowed exposure of the microbes to solar radiation, and effective disinfection was demonstrated as assessed by the reduction in *E. coli*. These kinds of SODIS systems are expected to have immense utility in regions near the equator and other tropical areas in developing countries where the intensity of UVB radiation is high (Mbonimpa et al., 2012).

6.2.8 Microbial Disinfection of Seawater: Ballast Water Management

Ballast water is the water that is pumped into the ship for its stability and manoeuvrability during its voyage. Water is taken in at one port, when the cargo is unloaded, and discharged at another port, when the cargo is loaded. The ballast water tanks are teeming with a variety of marine life. While most organisms will not endure the voyage to the port of discharge, some may survive and thrive in the new environment. These nonindigenous species (bioinvaders), if they become established, can wreak serious ecological and economic havoc on the new environment and on the health of humans. The adverse effects of introduction of nonnative species were shown by the discovery in the 1980s of the fouling European zebra mussel in the Great Lakes, a toxic Japanese dinoflagellate in Australia, and a carnivorous North American comb jellyfish in the Black Sea. These three introductions alone have cost many millions in remedial action and have had deep and ecological repercussions.

Publication of *Stemming the Tide* in 1996 fuelled research and development of ballast water treatment technologies. The International Maritime Organisation (IMO) also recognised the need for on-board ballast water treatment systems and, in February 2004, adopted the International Convention for the Control and Management of Ships' Ballast Water and Sediments, to regulate discharges of ballast water and reduce the risk of introducing nonnative species from ships' ballast water. The D-2 regulation of the convention sets the standard that ballast water treatment systems must meet (see Table 6.4).

This section will discuss the various technologies available for the treatment of ballast water. Most of these technologies are well established for the treatment of municipal and industrial water and have been discussed in detail earlier in the book. These need to be suitably modified for shipboard application. Changing ballast at sea is currently the technique to reduce the risk of introducing bioinvaders into the marine environment through discharged ballast water. An overall comparison of the technologies is presented in Table 6.5.

TABLE 6.4

IMO D2 Regulations

Serial No.	Type of Organism	Prescribed Limits
1	Intestinal enterococci	<100 CFU[a]/100 ml
2	*Escherichia coli*	<250 CFU[a]/100 ml
3	Toxicogenic *Vibrio cholerae*	<1 CFU[a]/100 ml
4	Plankton (10–50 µm)	<10 cells/ml
5	Plankton (>50 µm)	<10 cells/m³

Source: Data from Lloyd's Register, *Ballast Water Treatment Technology*, June 2011.

[a] Colony-forming unit.

6.2.8.1 Ballast Water Exchange

This method is usually effective because most freshwater or coastal organisms cannot survive when discharged into the ocean environment. Similarly, freshwater or coastal waters are inhospitable to oceanic organisms. Since the main function of ballast water is to ensure the stability of ships at sea, altering the ballast condition while on the journey may compromise vessel safety. Thus, changing ballast is not universally applicable and is not completely effective.

The effectiveness could be improved by redesigning ballast tanks and pumping systems. Continual flushing of the tank using multiple pipes, with one pipe bringing water into the tank and another pipe allowing water to exit the tank, would be safer because the tanks would contain water all the time (Chase et al., 2001). Also, loading and unloading of the water on high seas is unsafe for the structural integrity of the ship, and hence the practice is hazardous.

6.2.8.2 Ballast Water Treatment Technologies

6.2.8.2.1 Physical Separation Methods

For the removal of nonindigenous organisms from water, filtration and physical separation systems are environmentally benign, as they do not add any toxic substances (chemical biocides) to the ballast water.

6.2.8.2.1.1 Filtration   The separation and removal of organisms above a certain size from ballast water could be achieved using a shipboard filtration system. If filtration is done prior to intake of ballast water, the requirement of subsequent ballast change or further on-board treatment will be much less, and the filtered organisms may be retained in their native habitat (Chase et al., 2007). The options for on-board filtration systems are either mesh strainers (screens) or deep media filtration.

6.2.8.2.1.2 Strainers or Screens   A screen filter is a rigid or flexible screen to separate particles out of water. Strainers are screens that prevent the intake of large objects and organisms. For smaller filter openings (about 10 μm), the fouling and choking of the filter is more. Another disadvantage of strainers is that as the material deposits on the screen, the flow rate of water through the mesh decreases rapidly. Normally, this would have been compensated by using large areas of screens, but in shipboard applications there are space constraints. So, periodic cleaning, using automatic back flush systems, or scraping the systems needs to be carried out. Strainers are relatively compact, simple to operate, and amenable to retrofit on existing vessels, but they have limited effectiveness because many of the target organisms are smaller than the strainer size and can pass through the treatment system (National Research Council, 1996).

6.2.8.2.1.3 Media Filters   A media filter uses a bed of a certain type of material for filtration. Media filters can remove particles as small as 1 to 5 μm in diameter and most of the sediment load in the ballast water. They are reliable and would not pose any severe problems while retrofitting (National Research Council, 1996). Compressible media (e.g., crumb rubber filters, made from waste tires) may be suitable for potential shipboard installation due to its relatively compact size, which would require less space than conventional media filters (Kazumi, 2007).

6.2.8.2.1.4 Crumb Rubber Filtration Technology   Tang et al. (2005) studied an innovative crumb rubber filtration technology using discarded vehicle tires as the filter media. Due to its elasticity, the crumb rubber filter allows higher filtration rate, lower head loss, longer filtration run time, and better effluent quality.

   For a ship with a ballast flow rate of 5000 m^3/h, the crumb rubber filtration would require a surface area of 70 m^2. This size is much smaller than a typical sand filter. But it is too large for shipboard applications. A pilot test conducted by Tang et al. (2005) indicates that crumb rubber filtration removed up to 70% of phytoplankton and 45% of zooplankton, showing that crumb rubber filtration alone is not effective enough to meet the performance standards set by the International Maritime Organisation. So, it must be combined with another technology for effective disinfection of organisms.

6.2.8.2.2 Chemical Disinfection

There are chemical and physical means of disinfection of ballast water. Chemical treatment involves the use of biocides. Biocides are among the most widely used industrial chemicals, and there are a lot of data on their use in wastewater treatment. The concentrations of chemicals needed to treat ballast water are small. Simple online chemical injection pumps with the main ballast pumps can be used to add the biocide at regular intervals. The turbulence would ensure complete mixing of the biocide. However, the size should be considered while retrofitting vessels, care must be taken while handling hazardous substances, and the possibility of corrosion should be taken into account. The various oxidising and nonoxidising biocides have been discussed in detail in earlier chapters of this book, and here we just list the studies reported by researchers, when working with sea or brackish water.

6.2.8.2.2.1 Oxidising Biocides Chlorination (Electrochlorination) This technique is ideally suited for shipboard applications using the salinity of seawater itself. In this system, the main ballast flow is first filtered, and then a stream of 1% of the total ballast water volume is diverted to the electrolyser, where chlorine is formed at the anode and sodium hydroxide at the cathode, which react to form NaOCl and hydrogen gas. The hydrogen gas is diluted with air and vented to the atmosphere, removing any potential for hazard. The stream of water is then reintroduced to the remaining 99% main ballast flow. NaOCl dissociates to form OCl⁻ and HOCl (hypochlorous acid), which act as biocide disinfectants, as described earlier. Hypochlorite generation is therefore entirely *in situ* and is only done during the ballasting intake process. Some residual disinfectant that remains in the ship's ballast tanks during voyage prevents possible regrowth of organisms. However, as per the D_2 standards involving active surfaces, a subsequent treatment involving sodium thiosulphate is needed to tackle this residual disinfectant during the subsequent discharge of this treated ballast water back into the sea (Nielson, 2006).

$$|NaCl + H_2O \rightarrow NaOCl + H_2|$$

6.2.8.2.2.1.1 Efficacy and Safety
Nanayakkara et al. (2011) prepared an environment that simulated the ballast water using seawater and the organisms of interest. Disinfection performance of the electrochemical reactor was evaluated by inactivation of *E. coli* and *E. faecalis*, two microorganisms regulated by the IMO. It was found that the discharge standard of *E. coli* and *E. faecalis* can be met if energy consumption is as low as 0.006 KWh/m³. Other organisms would be killed by residual oxidant. The chlorine disappeared completely in 3 days. Ecotoxicity tests were carried out to see the effect of treated ballast water on organisms that are known to be sensitive to toxins and concluded that the effect of treated ballast water on a new environment was insignificant.

6.2.8.2.2.2 Ozone
6.2.8.2.2.2.1 Principle of Operation
Ozonation involves ozone formation by passing oxygen-enriched compressed air between a series of water-cooled electrodes discharging a high-voltage (10,000 V) electric arc or corona. This converts a portion of the oxygen present into ozone, which is introduced into the ballast water stream via a Venturi injector (Figure 6.3) (Wright et al., 2010). Earlier in the chapter, the use of plasma and pulsed plasma were discussed.

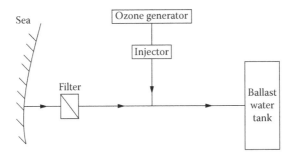

FIGURE 6.3
Ozonation of ballast water.

Most of the gas dissolves in water, decomposes, and reacts with other chemicals in the ballast water to kill organisms, but unlike chlorine, it does not have residual ozone, so there is no long-lasting protection.

Combining ozone with a treatment method that successfully eliminates larger organisms would be more effective than using ozone as a solitary treatment.

Most ozonation suppliers use an ozone dose of 1 to 2 mg/l, which has proven to be effective (Lloyd's Register, 2007).

6.2.8.2.2.2.2 Efficacy and Safety

Oemcke and Leeuwen (2005) investigated ozone for its potential to remove marine dinoflagellate algae from ships' ballast water by conducting lab-scale experiments. Dinoflagellate algae, *Amphidinium* sp., were used as indicators because these produce a type of cyst that is difficult to inactivate, but relatively easy to culture.

They concluded that high doses of ozone (5 to 11 mg/l) and up to 6 h of residual contact (ship voyage exceeding this time) were required for inactivation.

6.2.8.2.2.3 *Hydrogen Peroxide*
6.2.8.2.2.3.1 Principle of Operation

Hydrogen peroxide can be added in the form of the commercially available solution having a concentration of 25 to 35%, it can be added in the form of a compound supplying hydrogen peroxide or a compound that produces H_2O_2 in water-like organic peracids, or it can be produced by electrochemical decomposition of water or alkaline solution. The concentration of H_2O_2 required is about 10 to 500 ppm for disinfection of ballast water (Wakao et al., 2006).

6.2.8.2.2.3.2 Efficacy and Safety

Kuzirian et al. (2001) conducted tests on the use of hydrogen peroxide as a disinfectant and found that a concentration of 1 ppm hydrogen peroxide can be lethal to plankton composed of a wide mix of species, and a concentration of 3 ppm peroxide has effects comparable to ozone levels when tested on larvae of the mollusc *Hermissenda crassicornis.*

6.2.8.2.2.4 Nonoxidising Biocides   The nonoxidising biocides have an advantage over oxidising biocides because of the limited reactions of nonoxidising biocides with compounds present in ballast water. The residuals generated by nonoxidising biocides decay fairly rapidly and form nontoxic byproducts. Thus, treated water might not pose a substantial environmental hazard if it were discharged in large quantities (National Research Council, 1996). Also, the nonoxidising nature does not pose corrosion problems in ballast water tanks.

6.2.8.2.2.5 Glutaraldehyde   Glutaraldehyde is a five-carbon dialdehyde with the formula $CH_2(CH_2CHO)_2$. It is a pungent, colourless liquid commonly used as a medical disinfectant. Glutaraldehyde's biocidal properties are attributed primarily to the reaction of the aldehyde group with proteins in the organisms, and hence destroy them (Sano et al., 2003).

6.2.8.2.2.6 Vitamin K3 (Menadione)   Menadione, or vitamin K3, is an unusual chemical for treatment, as it is a natural product (but it is produced synthetically for commercial use). It is relatively safe to handle. It is marketed for use in ballast water treatment under the proprietary name SeaKleen by Hyde Marine.

   6.2.8.2.2.6.1 Efficacy
Experiments by Gregg and and Gustaaf (2007) to test the efficacy of SeaKleen in ballast water treatment showed that it eliminated vegetative microalgae at 2 ppm concentration and could control cysts of dinoflagellates *Gymnodinium catenatum* and *Protoceratium reticulatum* at concentrations of 6 and 10 ppm respectively, when exposed for a period of 2 weeks, but it could not kill the dinoflagellate *Alexandrium catenella.*

6.2.8.2.3 Physical Disinfection
6.2.8.2.3.1 Deoxygenation
   6.2.8.2.3.1.1 Principle of Operation
Deoxygenation involves stripping of the oxygen from the ballast water, which will eliminate many potential bioinvaders due to lack of oxygen. Some organisms that require oxygen can survive short periods of anoxia (Tamburri, Wasson, and Matsuda, 2002), but they are usually inactive under such conditions. Oxygen can be removed by pumping with an inert gas (McNulty, 2003), using a vacuum system (Browning, 2001), or by biological deoxygenation (Blois, 2008). The ballast water could also be combined with carbon dioxide (from the combustion products from the engine) to displace the dissolved oxygen (Lloyd's Register, 2007). The nitrogen pumping method is particularly attractive when a supply of nitrogen gas exists on board (air-based pressure swing adsorption systems), as when transporting flammable or explosive materials such as liquid fuel. It also has the added advantage over other processes that it prevents corrosion inside the ballast tanks.

6.2.8.2.3.1.2 Efficacy and Safety

Deoxygenation is toxic to a range of fish, invertebrate larvae, and aerobic bacteria but is ineffective against anaerobic bacteria, cysts (dinoflagellate cysts), and would provide only a partial solution to eliminating the bioinvaders.

To destroy most organisms, ballast water would need to be heated to temperatures in the range of 35 to 45°C and held there for a set period of time. Some voyages will be too short to permit heating the water to the required temperature. The volume of ballast water that can be treated is also limited by the amount of energy available from waste heat sources.

It is difficult to achieve uniform heating rates (Boldor et al., 2008). Higher organisms, such as fish, are more easily killed by thermal treatment than are the microbes.

6.2.8.2.3.2 *Pulsed Plasma and Electrical Pulse Technology*
6.2.8.2.3.2.1 Principle of Operation

In pulsed electrical field technology and pulse plasma technology, short bursts of energy are used to kill microorganisms in water by destroying their cell membranes (Ryu et al., 2010). The only difference between the two is the way in which the energy is created. Neither produces toxic chemical residuals, but a pulsed system may generate gaseous decomposition products, particularly carbon dioxide. The systems are automated and do not require attended operation. The treatment is sensitive to the presence of sediment. So, filtration may be required prior to treatment. The concept and its variation have been discussed earlier in detail. Figure 6.4 indicates a possible working scheme using pulsed plasma.

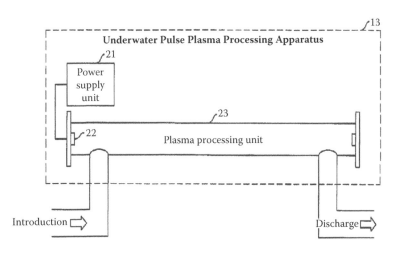

FIGURE 6.4
Plasma processing unit. (From Ryu, H.S., et al., U.S. Patent 2010/0126940, 2010.)

6.2.8.2.3.2.2 Efficacy and Safety

Using the electric pulse system, 95 to 99.9% sterilisation of brine shrimp, *Artemia salina*, is claimed by the National Research Council (1996). The electrical energy required for this is about 1 J/cm^3 for pulse duration of 3×10^{-5} s. using about 600 kW of power.

6.2.8.2.3.3 Ultraviolet Treatment
6.2.8.2.3.3.1 Principle of Operation

This well-established technology uses ultraviolet energy from UV lamps (with a wavelength of about 200 nm) to inactivate microorganisms and acts by disrupting the DNA within cells, thereby prohibiting their replication (National Research Council, 1996). The details of the possible systems and their efficacy have been discussed in Chapter 3.

6.2.8.2.3.3.2 Efficacy and Safety

One of the main drawbacks of ultraviolet disinfection is that the efficiency is greatly reduced in water containing suspended matter. So, ballast water needs to be filtered before treatment. Also, it is not effective against higher organisms, cysts, and spores. Thus, it should be combined with another method to effectively remove all potential bioinvaders from ballast water (Chase et al., 2007).

6.2.8.2.3.4 Hydrodynamic and Acoustic Cavitation
6.2.8.2.3.4.1 Principle of Operation

This was discussed in detail in Chapter 5. In this subsection, we will see its possible application for ballast water treatment.

Cavitation is the phenomenon of formation, growth, and collapse of microbubbles within a liquid due to pressure variation in the flowing liquid. It can be produced by acoustic means (ultrasound) or by hydrodynamic means (flow-through constrictions).

Acoustic systems use transducers to apply sound energy of specified amplitude and frequency, which causes cavitation and results in localised mechanical stresses that disrupt cells.

Due to application of ultrasound, alternating compressions and rarefactions are generated in the liquid to be treated. The resulting cavitation is influenced by frequency, power density, time of exposure, and the physical and chemical properties of the liquid. Optimum frequencies for destroying microorganisms are reported to be from 15 to 100 kHz. Effectiveness decreases with increasing distance from the transducer as the energy density in the liquid decreases. The efficacy of ultrasonic treatment increases with exposure time and can also be influenced by resonance effects due to container geometry (National Research Council, 1996).

In hydrodynamic cavitation, a fluid is passed through a constriction. The velocity of the fluid increases and the static pressure decreases. As velocity

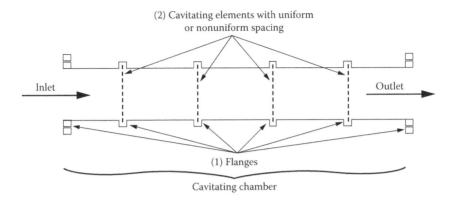

FIGURE 6.5
Hydrodynamic cavitation system. (From Bhalchandra, A.P., et al., U.S. Patent 7,815,810, 2010.)

increases, if pressure falls below a critical value (usually below the vapour pressure or the operating pressure), small bubbles or cavities are formed and a high-magnitude pressure pulse is created (of the order of hundreds of bars). This is enough to rupture microbial cells. Asymmetric cavities that collapse result in high-speed liquid jets that kill microbes due to shear. The apparatus consists of a cavitating chamber with single or multiple elements made of metal, ceramic, or plastic, of varying thickness, having single or multiple orifices, placed perpendicular to the direction of flow of liquid (Figure 6.5). The water may be recirculated for additional disinfection (Bhalchandra et al., 2010).

6.2.8.2.3.4.2 Efficacy and Safety
The ultrasonic system could totally destroy a broad spectrum of fungi, yeasts, and pathogenic bacteria. But pilot-scale ultrasonic treatment tests on zebra mussels, at flow rates between 0.19 and 1.89 m^3/h, showed that when treatment time was less than 60 s, kill rate was small, but it increased to 100% with a treatment time of 12 min. High-frequency sound (approximately 125 kHz) shows strong avoidance responses from certain fish. Ultrasound treatment penetration is small and performance deteriorates with scale-up (*Alosa pseudohargenus* and *Alosa sapidissima*), but the fish are not killed (Kazumi, 2007).

The apparatus is simple, eco-friendly, and can be retrofitted onto any vessel with minor modifications. However, the energy consumption is significant. Table 6.5 gives a comparison of all the possible techniques that are likely to be partially or totally successful in a variety of shipboard situations.

Thus, ship ballast tanks represent a unique challenge for treating water due to the high flow rates, large volumes, organism diversity, and ballast water residence time. From the survey of literature conducted, it appears that with the currently available technology, one overall on-board treatment method will not be applicable to all vessels carrying ballast water over the whole world. A combination of two or more treatments will effectively

TABLE 6.5

Comparison of Ballast Water Treatment Technologies

Process	Treatment Method	Benefits	Drawbacks	Capex Rs.(Lacs)/1000m³/h	Opex Rs/1000m³/h	Footprint m²/1000m³/h	Power Consumed kW/1000m³/h	Comments
Solid–Liquid Separation								
Filtration	Screens, strainers, crumb rubber technology	Can filter larger organisms and sediments	Needs periodic backwashing, cannot filter out small target organisms, needs large surface area			14		Needs to be used in conjunction with another disinfecting technology
Hydrocyclone	High-velocity centrifugal rotation of water to separate particles	No moving parts, easy operation	Less effective than filters, effective only for larger organisms					Effectiveness depends on density of particle and water, speed of rotation, size of particle
Chemical Disinfection (Oxidising Biocides)								
Chlorination	Biocide is dosed in water to destroy cell walls of organisms	Cl₂ gas is inexpensive, sodium hypochlorite can be injected into ballast water stream	Cl₂ gas is dangerous, organic matter in seawater forms toxic compounds on oxidation, residuals need to be removed					Only an additional pump is required as additional equipment, but large space to store the chemicals is needed

Method	Mechanism	Advantages	Disadvantages					Notes
Electrochlorination	Electrolytic decomposition of seawater to ⁻OCl and HOCl	Corrosion rates are insignificant, does not require storage of chemicals	Ineffective against cysts, forms harmful by-products	257.77	1000	8.31	52	Treated water can be released into discharge port; residual oxidants prevent regrowth of microorganisms
Ozonation	Bubbling ozone gas through ballast water containing Br⁻ to give OBr⁻ as the oxidising species	Effective at eliminating prokaryotic and eukaryotic organisms, results comply with IMO standards	Ineffective against large organisms, may form bromoform, which is a known carcinogen, ozone gas is toxic to humans			10.21	63	Can be used with a treatment that eliminates larger organisms; a dosage of 1–2 mg/l is effective; the size of machinery is relatively large
Hydrogen peroxide and peracetic acid	Kills microorganisms by disruption of their cell membrane, via the hydroxyl radical (HO·)	Reagent is infinitely soluble in water, produces few harmful by-products, and is relatively stable	Reagent is relatively expensive, requires considerable storage facilities, needs to store the treated effluent for at least 6 days for safe discharge		10,000	5.54	25	Concentration of H_2O_2 required is 10 to 500 ppm; this is a quite high dosage; rapid regrowth of microbes occurs when it is no longer present

(Continued)

TABLE 6.5 (CONTINUED)

Comparison of Ballast Water Treatment Technologies

Process	Treatment Method	Benefits	Drawbacks	Capex Rs.(Lacs)/1000m³/h	Opex Rs/1000m³/h	Footprint m²/1000m³/h	Power Consumed kW/1000m³/h	Comments
Chemical Disinfection (Nonoxidising Biocides)								
Glutaraldehyde	Reaction of the aldehyde group with proteins in the organisms, thus destroying them	Decays fairly rapidly, forms nontoxic by-products, no corrosion problems in tanks	Extremely high dosage is required		Rs.250 k per treatment (500 ppm at 200,000 kg)			500 ppm of glutaraldehyde held for 24 h kills 90% of organisms
Vitamin K (menadione)	It is toxic to invertebrates	It is a natural product, relatively safe to handle	It cannot kill certain dinoflagellates, slow rate of degradation, discharge of residual toxic water, poor bactericidal properties					Needs treatment before discharging; may not be suitable to use for ballast water treatment on a large scale
Physical Disinfection								
Deoxygenation	Stripping of the oxygen from ballast water by inert gas pumping, using vacuum, or biological means to asphyxiate organisms	Inert atmosphere prevents corrosion inside the ballast tanks, no safety issues, no explosion hazards, no discharge of toxic compounds	The vacuum mechanism is less efficient, not a good option for ships with short transit times	207.75		5.84		It represents a net savings for ship owners due to a decrease in corrosion; savings can be Rs.35(Lacs)/vessel/year when compared with painting technology

Method	Principle	Advantages					Disadvantages	Remarks
Thermal treatment	Heating by using heat from engines, pumping engine cooling water into ballast tanks, or using continuous microwave techniques	Waste heat utilisation from engines, no chemical by-products, minimal corrosion		8.92			Not useful for short voyages, limited volume of water can be treated, heated water damages discharge port environment	Filtering out dead organisms is needed; only additional piping and a pump will be required
Pulsed plasma and electrical pulse	Short bursts of energy (electrical or plasma pulse) are used to kill microorganisms by destroying their cell membranes	No toxic chemical residuals, automated systems do not require attended operation			720	175	Due to release of soft x-rays and high-energy UV radiation, the system should be properly shielded and encapsulated	Filtration is required prior to treatment
Ultraviolet	Amalgam lamps surrounded by quartz sleeves create ultraviolet energy to denature the DNA of microorganisms, thus preventing reproduction	No residuals are formed	111	9.23			Not effective against higher organisms, cysts, and spores	Prior filtration is necessary as ballast water should be clear; it should be combined with another method to remove all potential bioinvaders
Hydrodynamic and acoustic cavitation	Applying ultrasonic energy or passing fluid through a constriction causes cavitation, which disrupts cells	No health risk to crew, does not require special skills or additional manpower, no toxic by-products	20.5	4.54	2500	46.16	Not effective against higher organisms like certain fish; ultrasonic treatment penetration is small and performance deteriorates with scale-up	Optimum frequencies for destroying microorganisms are reported to be from 15 to 100 kHz; the efficacy of treatment increases with exposure time

remove target organisms. While selecting a treatment method, one must consider safety of the ship and the crew, the effectiveness of removing the target organism, the ease of operation of equipment, the cost of treatment, the footprint, and the compliance with standards.

Ultraviolet treatment needs water free from sediments. Some treatments are not effective against higher organisms (e.g., ultraviolet treatment and ozonation). For these scenarios, physical separation methods must be combined with the disinfection technology. The most commonly used separation technique uses screens or strainers, but the new crumb rubber media filter technology, which uses waste tires, is also effective. Physical methods of disinfection (electric and pulse plasma, and acoustic and hydrodynamic cavitation) are also gaining popularity due to the absence of chemical residues and high efficacy. But, these technologies are quite costly and need to be further developed. Thermal treatment is practical only for certain vessels for long voyages and has a limited capacity. Deoxygenation, although not completely effective in eliminating all target organisms, has proven to decrease corrosion rates in ships, substantially giving it an added economic advantage for ship owners.

Taking into account various factors, the chemical disinfection method seems to be a promising option for treating ballast water in conjunction with filtration. Considerable data are available on this technology, as it is the most widely used for the treatment of wastewater, and hence the most economical and effective means to carry out disinfection have already been established. The dangers of forming toxic organic compounds (DBPs or disinfection by-products) can be avoided by prior filtration and removal of sediments. Also, if electrochemical methods are used, storage of chemicals on board will not be required, thus decreasing the footprint and need for additional manpower for dosage.

Currently, ballast water treatment is not widely practiced, as installing new technologies or retrofitting ships is expensive. Ship owners are reluctant to use a new technology unless it is proven effective. Thus, preventing release of nonnative species through ballast water remains a low priority, and many of the treatment options are still in the experimental stage. The establishment of more stringent regulations and implementation of changes in ballast water management will bring about the accelerated development of ballast water treatment technologies.

6.2.9 Supercritical Water Oxidation (SCWO)

As the name suggests, this transformation is carried out at the condition beyond the critical point of water (P_c = 220.55 bar and T_c = 373.98°C). This is considered an extension of the process called wet-air oxidation or flameless combustion or oxidation. The process is also sometimes called hydrothermal oxidation (HTO).

This process of the treatment of water involves the water being heated above the critical temperature and using pressures in excess of that generated autogenously (generated by the water vapour). The oxidation requires the supply of oxidising agent, which could be in liquid form (such as hydrogen peroxide) or would be in liquid form (such as air or pure oxygen). At the critical point of water, the latent heat of vapourisation is zero (against 540 Kcal/kg of water at atmospheric pressure and temperature of 100°C), and hence the transfer of the oxidising species; that is, oxygen from a liquid oxidising agent or from air does not experience any interphase mass transfer limitations/resistances. This renders the process very fast and quick and is essentially controlled by the oxidation kinetics of the pollutants on the materials to be oxidised, which is indeed very fast.

In the basic treatment system, aqueous mass (effluent or water containing organic, inorganic, and biological pollutants) is combined with an oxidiser at elevated pressure and temperature ($P > 221$ bar and $T > 550$°C) in a reactor, where the average residence time is only a few seconds (5 to 15 s) but always less than a minute.

Since the operating conditions are above the critical point, the entire mass exists as a homogeneous single phase, eliminating the interphase mass transfer resistances, completing the reaction to the extent of 99.99%. The completion of the reaction results in fully stabilised oxidation products (complete mineralisation to CO_2, H_2O, NO_X, etc.). Once the mineralisation is complete (decided by the kinetics of the oxidation reaction and the allowed residence time), the effluent is cooled and depressurised and separated into noncondensable gases and liquid phases. The process until this point is completely closed to the surrounding still, and hence is amenable to monitoring for assessing the targeted mineralisation efficiency. Figure 6.6 shows the schematic of the SCWO process, which is not limited to the destruction and mineralisation of the simple and complex organic and inorganic pollutants, but can also handle nitrates, ammonia, and various microbiological pollutants most efficiently with a proper choice of the oxidising or reducing agent. Since most of these reactions are oxidising and essentially exothermic in nature, the overall energy balance indicates that once the operating conditions are reached using external fuel at a contamination level of 10,000 ppm of chemical oxygen demand (COD) or above, the process can be completely self-sustainable, requiring no additional external energy input. At pollutant concentrations above 10,000 ppm of COD, the process could be a net energy producer, as also indicated in Figure 6.6.

6.2.9.1 Chemical Reactions

The following set of chemical reactions is known to occur (www.turbosynthesis.com):

Material	SCWO Reaction	
Cellulose	$C_6H_{10}O_5 + 6O_2 \rightarrow 6CO_2 + 5H_2O$	
Methane	$CH_4 + 2O_2 \rightarrow CO_2 + 2H_2O$	
Benzene	$C_6 + H_6 + 7.5O_2 \rightarrow 6CO_2 + 3H_2O.$	
DOXIN	$Cl_2\text{-}C_6H_2\text{-}O_2\text{-}C_6H_2\text{-}Cl^2 + 11O_2$	$12CO_2 + 4\ HCl$
TNT	$CH_3\text{-}C_6H_2\text{-}(NO_2)_3 + 5.25O_2$	$7CO_2 + 2.5H_2O + 1.5N_2$
Ferrous chloride	$FeCl_2 + 0.25O_2 + H_2O$	$0.5Fe_2O_3 + 2HCl$

As can be seen from the above, the process of halogenated or sulphur-bearing compounds is expected to generate respective acids as the final product. The material of construction (MOC) of the reactor should withstand these chemicals at such extreme conditions for the safety of the operation.

6.2.9.2 Engineering Challenges

Even though the basic chemistry is simple and straightforward, implementation of the same at such extreme operating conditions and at such large scale of operation is difficult. Experience has shown that the corrosion rates could be significant, especially in the presence of halogenated pollutants, and even exotic materials such as Hastealloy C-276 or Inconel 625 do not provide adequate protection and reliability. Recent development of titanium-lined autoclaves has offered a partial solution to the corrosion problem and a temperature up to 650°C could be handled.

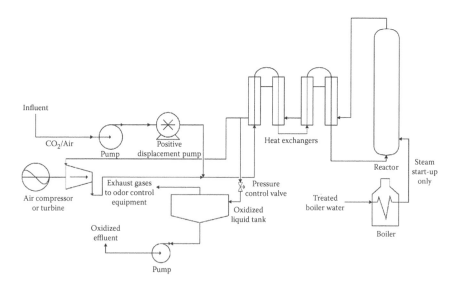

FIGURE 6.6
Schematic of the SCWO process.

Even though the solubility of gases and oxidising liquids in the aqueous effluent increases with the operating pressure, the solubility of the salts (either present initially or formed as a product) reduces drastically as one approaches the critical point. The handling of this precipitated salt could be a major engineering bottleneck in a continuous treatment scheme using SCWO, as plugging and choking could render the entire process unsafe.

In the traditional SCWO or wet air oxidation (WAO) process the entire high-temperature, high-pressure stream is depressurised (use of turbines to run the high-pressure pumps to pressurise incoming effluent upon exchanging its enthalpy with the incoming effluent has also shown considerable energy integration potential) and cooled. The treated water can be reused. Use of the closed-cycle SCWO process or the transpiring wall SCWO process (www.turbosynthesis.com) is among the recently introduced innovations to tackle this solid handling problem using interphase mass transfer or direct mixing with incoming effluent for dilution, respectively.

6.2.9.3 Operating Range for SCWO Processing

As the critical pressure is in excess of 220 bar, traversing a supercritical isotherm from, say, 250 bar to lower pressures, the corresponding changes in the physicochemical properties of the influent and the by-products are only marginal. This suggests that the continuum of reactor operating points may exist along any supercritical isotherm. Work on methane oxidation reported by Sandia National Laboratory (1996) indicates 135 bar operating pressures are significantly better than 270 bar. The proposed diffusion type hydrothermal reactor (also called transpiring wall SCWO reactor) works with operating pressures as low as 15 bar. Table 6.6 shows the changes in the activation energy, order of reaction, and other kinetic parameters as a function of the operating pressures of the SCWO process.

TABLE 6.6

Changes in the Activation Energy, Order of Reaction, and Other Kinetic Parameters as a Function of the Operating Pressures of the SCWO Process

Serial No.	Parameter	Units	Value I	Value II
1	Reactor pressure	Bar	135.00	270.00
2	Preexponential factor, log A	gmol, l, s	35.00	24.00
3	Activation energy, Ea	Kcal/gmol	115.00	80.90
4	Reaction order w.r.t. CH_4	—	1.80	1.74
5	Reaction order w.r.t. O_2	—	0.15	1.24
6	Variance	—	3.00E-05	8.80E-05

Source: Data from www.turbosynthesis.com.

6.2.9.4 Specific Research Works

Discussed below are the specific works in the field of SCWO, practiced and experimented with for lab and field application.

Perrut (2011) reports a summary of the present state of the art, underlining the promising future of supercritical fluid application for sterilisation/pasteurisation and inactivation of virus as an alternative green method to the conventional process that can be used for many thermolabile products due to the heat degradation during sterilisation. A comparison of SCWO or supercritical fluids (SCFs) with other sterilisation techniques, such as the use of sterilisation chemicals (H_2O_2, ethylene oxide, peracetic acid, etc.) or radiolysis/γ-ray irradiation, has pointed out the opportunities and challenges.

One of the outstanding pieces of work was recently published by Xu and coworkers (2012). They have studied all the engineering parameters required for the designing and running of the SCWO plant for the destruction of the sewage sludge, similar to the attempts made in Germany, where sewage sludge is used as a feed for anaerobic digester for biogas production. Xu and coworkers have designed and optimised a pilot plant trying to find a plausible solution to the problems of corrosion, plugging, and high running costs of the SCWO process in a Chinese context. A simple comparison with incineration as an alternative has been carried out, indicating the problem areas and the work still needed to come up with a working engineering solution. The work clearly indicated the efficacy of SCWO and its economical attractiveness as an alternative. As discussed by Xu et al., even though for a simple sewage sludge this indeed may work and could be an alternative to incineration, incineration offers an energy recovery option readily, whereas considerable plant configurational modifications are required in the case of SCWO for energy recovery, and it may become an expensive proposition. Similarly, for more complex effluents containing halogenated pollutants, the problems related to the MOC of the reactor and heat transfer equipment still remain unresolved.

Along similar lines Anikeev and coworkers (n.d.) have developed a mobile SCW plant in the Russian context. They have successfully destroyed with greater than 99% efficiency organic pollutants such as ketones and aldehydes, using H_2O_2, or air as oxidants, or also nitrate as an inorganic oxidant. The concept of development of a mobile unit was with an intention of on-site quick feasibility assessment of the variety of wastes and establishment of the design and operating parameters for the development of subsequent optimised design for field application.

They have also described the role of a catalyst such as MnO_2, when used in conjunction with SCWO, in reducing the operating temperature and pressure requirements, significantly reducing the material cost.

Sharma and coworkers (2012) have written a review specifically addressing the issue of cynotoxins, particularly microcystins (MCs) spread in the environment arising out of cyanobacteria blooms and their destruction using advanced oxidation processes with chemical oxidants such as O_3, H_2O_2, permanganate, and other chlorine-based oxidants. They have discussed the destruction of MCs using advanced oxidation processes and compared it with that of conventional oxidation processes, highlighting the degree of degradation, reaction kinetics and its possible degradation pathways, and toxicity of the reactants and degradation by product.

Thus, in conclusion of this section, one can say that the thermal treatment, such as WAO and SCWO (principally both are the same; they differ in terms of their operating parametric ranges), could be an attractive treatment method, technically and reactively simple in terms of chemistry for recycle and reuse of water associated with the aqueous effluent. However, considerable engineering challenges and higher operating cost issues need to be resolved before it can be universally accepted. Also, the technological sophistication of the control strategy needs to be fail-safe due to the extreme operating conditions to increase its acceptability in a relatively low-tech water treatment industry.

6.2.10 Membrane-Based Treatment

All over the world, a realisation has been felt that membrane-based processes are likely to play an increasingly significant role in water treatment and management. The earlier scepticism and pilot-scale studies in the 1970s have given way to full-scale industrial water recovery systems involving micro, nano/ultra, and reverse osmosis membranes. Considerable developmental work has occurred in the industry addressing the new membrane materials (in addition to various polymeric membranes), such as ceramic membranes, capable of withstanding higher temperature, giving rise to hybrid processes combining thermal treatment and filtration in one go. Engineering efforts have also been made, looking specifically into the area of increasing the life and reliability of the membrane through appropriate pretreatment, that is, microfiltration before reverse osmosis or by recovering the pressure energy of the outgoing stream by a clever engineering configuration of the system.

The basics of the membrane-based filtration systems involving micro, ultra/nano, and reverse osmosis techniques, their hardware configuration, separation of the suspended particulate matter, including microbial contamination, using micro-, ultra-, and nanofiltration membranes, respectively, and the dissolved solutes, using reverse osmosis, their working principles, problems associated with fouling and concentration polarisation, and so forth, have been the subject of many monographs and books (Wang et al., 2011; Van Rijn, 2004; Baker, 2000). In this section, we will concentrate on the recent development, where multiple operations are at work simultaneously and efforts of the researcher's work which is in that direction.

One of the major developments of the last decade is the membrane bioreactor (MBR), where the biological treatment of the wastewater is combined with membrane for *in situ* separation and reclamation of the treated water. The membrane module consisting of micro- and ultrafiltration units can be either externally linked to the aerobic bioreactor or submerged, providing an effective barrier for the biomass and other organic (large molecular weight) pollutants from entering the treated and unfiltered water. The water coming/reclaimed from MBR can be directly used as a process water, as it can be fed to the reverse osmosis membrane, giving potable quality water devoid of salts and microbes/viruses.

MBR offers a number of advantages over conventional activated sludge processes, such as a smaller footprint and better quality of treated water. The MBR system overcomes the usual limitation on mother liquor volatile suspended solids (MLVSS) concentrations in a conventional activated sludge process, allowing it to operate with a significantly higher concentration of MLVSS, reducing the hydraulic retention time, and thereby reducing the volume and size of the bioreactor.

Two key challenges that still need to be addressed are:

1. Membrane fouling by microorganism: This challenge is currently handled by intense aeration (any way it is required to satisfy the increased respiration rate of higher MLVSS) to agitate and scour the membrane surface (be it hallow fibre or flat sheet). Also, an attempt to improve the antibiofouling efficiency of the polyether sulphone membrane by functionalising it with zwitterionic monomer has been made by Razi et al. (2012) with some success.

2. Increased energy requirement for the supply of required respiratory oxygen; this is currently being handled by combining the diffused air aeration system, strategically placing the gas diffuses, giving values of kg of O_2 transferred/KW.h in excess of 2.5 to 3.0 (conventional aerators give these values in the range of 1 to 2).

An excellent treatise on the subject can be found out in an article by Hu et al. (2012). They proposed that the concepts of membrane distillation (also sometimes referred to as pervaporation) and forward osmosis (FO) are emerging areas, which are likely to have a significant impact on the energy consumption in the field of desalination and water reuse. FO is comparable to reverse osmosis (RO) in terms of its operation principles, except in the range of the solute concentration difference and subsequent lowering of driving pressure gradient requiring lower energy levels. A pervaporisation membrane capable of giving water flux comparable to that of RO membranes is still in the development stage.

Microbial contamination in recovered/reclaimed water is a major concern. According to Hu et al. (2012), based on the work carried out at National University of Singapore (NUS), RO polyamide membrane provided the most effective removal MS2 bacteriophage (a virus-infecting bacteria). They concluded that size exclusion and change repulsions are the two dominant mechanisms for the removal of the MS2 virus. The performance seems to improve with the usage, which was attributed to the irreversible fouling giving additional resistance (cake resistance in addition to the intrinsic membrane resistance).

Other notable recent developments can be listed as the work by Guo and Hu (2012), where they combined coagulation and membrane filtration to pretreat the turbid water for successful subsequent UV treatment. UV treatment is known to get affected by the turbidity due to the reduced UV ray penetration, and hence the treatment of the water to remove or reduce the turbidity is an important pretreatment step. Interestingly, they found that the inactivation of MS2 bacteriaphage was more dominant due to coagulation than UV treatment; however, it is an interesting concept, where two-stage treatment can show some synergy.

By manipulating the timing and intermittency of operation in an ultra-low-pressure ultrafiltration system, Peter Varbanaets et al. (2010) have shown that for a household operation, 21 h/filtration operation and 3 h standstill resulted in an optimum flux without compromising the quality of water or the membrane permeability. The standstill allowed the fouling layer to relax and expand, resulting in an increased overall daily flux, suitable for household operation.

Another novel attempt reported by Zhang and coworkers (2012) involved the preparation of CTS/PVP composite film, which was *in situ* loaded with TiO_2 and Ag. The objective of this work was to prepare an antibacterial composite, which could be used for the storage of potable water. Based on Fourier transform–infrared (FT-IR) studies, they concluded that NH and –OH of CTS have some interaction with Ag nanoparticles, where the later were reduced *in situ*, increasing and sustaining its bactericidal activity, and were safe to use over an extended period.

Sanaeepur et al. (2012) has described the use of extractive membrane bioreactor for denitrification (nitrate extraction from drinking water) and have successfully modelled the system, explaining the working hydrodynamics of MBR. The approach is likely to be very useful for explaining the flux of the similar target impurities across the specified membranes, facilitating its optimised design.

The approach chosen by Chu et al. (2012) by immobilising the diatomite on a stainless steel wire mesh in developing the biodiatomite dynamic membrane bioreactor (BDDMBR) is interesting. The layer of diatomite that acts as a microorganism carrier and forms a layer on a stainless steel wire mesh acts as a fixed bed for the removal of organic matter and ammonical nitrogen. The degradation of the pollutants due to microbial action was slow, and hence, even though its filtration efficiency was good, the pollutant removal rate was unacceptable for it to be used as the potable water.

An excellent review on hybrid membrane processes using activated carbon pretreatment and other pretreatments has been published by Stoquart et al. (2012). They have explored a number of adsorptive and filtration processes in different sequences and have evaluated their performance to produce drinking water. The problems of membrane fouling and erosion have been quantified and remedial measures have been suggested.

Thus, in conclusion, it can be said that membranes and membrane bioreactors show considerable promise in the areas of water treatment, effluent treatment, water recovery, and reuse.

6.2.11 Photocatalysis for Water Treatment

To understand the working principle and its possible applications, the reader is referred to an outstanding treatise by Hashimoto and coworkers (2007). Photolysis and photocatalysis are among the advanced oxidation processes, relying mostly on the generation and attack of OH^- radicals, one of the strongest oxidising agents. These generated OH^- radicals can attack any type of oxidising pollutants as long as they are in the vicinity, as the half-life of these OH^- radicals is very short. Thus, the pollutant species need to be necessarily adsorbed on the photocatalyst, where these OH^- radicals are generated using UV radiations.

Hashimoto and coworkers traced the historical development of this area starting with the 1950s, where the action of UV and solar radiations on the TiO_2 surface was first observed and attempts were made to explain the possible mechanism. In the 1960s, photoelectrolysis of water was attempted using a single-crystal H-type TiO_2 (rutile) semiconductor electrode, and solar photoelectrolysis was demonstrated in the late 1960s. During the first crude oil crisis of the early 1970s, the research received a major impetus due to the possibility of oxidation at the TiO_2 electrode and reduction (hydrogen generation) at the Pt electrode. Many other semiconductor materials as electrodes, such as $SrTiO_3$, $KTaO_3$, and ZnO_5, were tried along with CdS and CdSe (for their small band gap). However, their efficiency and stability were much lower than those of TiO_2.

The other approach used by the researchers was to extend the utility of the wavelength of the solar spectrum to visible range (>415 nm) by adding photosensitiser dyes to the basic TiO_2 matrix. However, again the chemical and structural stability of these dyes was a major question. In the 1990s, O'Regan and Grätzel (1991) showed that porous TiO_2 electrodes and ruthenium complexes worked well with solar conversion efficiencies in excess of 10 to 11% and also successfully addressed the problem of stability.

TiO_2 as a photocatalyst in the powder form was tried in the early 1980s. It was realised that a powder system can never be used for water splitting to generate hydrogen and oxygen due to the fast recombination reaction; however, OH^- can be scavenged by organic moieties such as alcohol,

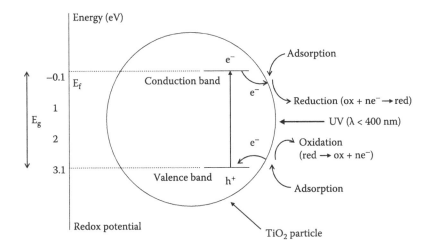

FIGURE 6.7
Typical scheme of photocatalytic process over TiO_2.

and hydrogen generation efficiency can be substantially enhanced. This approach, however, was the starting point of the research in the area of pollutant mineralisation using photocatalysis. The first report by Frank and Bard (1977) indicated the mineralisation and decomposition of cyanide using aqueous TiO_2 suspension.

In principle, both the reduction and oxidation sites are located on the TiO_2 surface, and the reduction of the adsorbed oxygen molecule (from O_2 dissolved in the water) proceeds on the TiO_2 surface. The holes generated by the UV radiation (h+) on the TiO_2 surface are highly oxidising, and most of the organic molecules are completely oxidised to their final stable oxidation state. Figure 6.7 shows a schematic representation of the photocatalytic phenomena. Thus, the focus shifted from hydrogen production to the destruction (oxidation) of a variety of pollutants, which was carried out quite successfully. Similar to SCWO, most of the pollutants would be quite readily mineralised; however, the rates were significantly slower.

The research and work reported can be broadly classified in terms of (1) reactor configuration, based on the suspended catalyst or immobilised catalyst or the type of the radiation source (solar, UV lamp, etc.) and the scale of operation, that is, effluent treatment for discharge or treatment for drinking water, and (2) types of pollutants that have been successfully mineralised, that is, organic, microbial, and so forth.

6.2.11.1 Types of Reactor Configurations and Their Hybridisation

Jin et al. (2012) have reported studies on immobilised TiO_2 film on glass substrate, irradiated with solar radiations, as well as with UV lamps. The water was filtered to remove the turbidity and was allowed to flow on the glass substrate coated with nano-TiO_2 film. They concluded that the supported TiO_2 thin film had good removal efficiency for trace organic pollutants reaching 45 to 63% within 2 to 3 h. He also concluded that, in an increasing order of importance, light intensity > dissolved oxygen concentration > reaction time > pH were the main influencing parameters. Organic pollutants were completely mineralised and hv degradation by-products (DBPs) were readily detected.

Along similar lines is the work on photolysis (without the catalyst) by Mbonimpa and coworkers (2012) using compound. Parabolic solar radiation concentration (CPC) for concentrating and focussing on glass tubes, through which the water was made to flow, is also noteworthy. Concentrating the UVB radiation had substantial germicidal effect for the inactivation of *E. coli*, and a DNA-weighted UVB dose was found to be the governing parameter. The reactor configuration showed linear stability and could be easily employed at a small community level for safe potable drinking water, or in disaster-affected areas. Even though disinfection occurred, whether it is due to concentration of UVB radiation intensity or increased temperature is not clear from their work.

Papageorgio and coworkers (2012) proposed a novel substrate for the immobilisation of TiO_2 and calcium alginate fibres. The aerogel of alginate fibres, prepared by supercritical CO_2 drying, loaded with TiO_2, showed excellent porosity and flow characteristics. Composite alginate/photocatalyst porous fibres were combined with TiO_2 membranes in a continuous-flow, hybrid photocatalyic/ultrafiltration water treatment process, resulting in a threefold enhancement in the target methyl orange removal efficiency. The proposed configuration shows a considerable merit and scale-up possibility. Similarly, Liu and coworkers (2012) showed the efficacy of the Ag/TiO_2 nanofibre membrane, fabricated using the electrospinning technique for bacterial degradation (>99.9% inactivation). The synthesised membrane also handled successfully the removal/mineralisation of an organic dye with more than 80% efficacy.

Sontakke et al. (2012) showed the utility of combustion synthesised TiO_2 in a suspended form for the inactivation of *E. coli*. They concluded that the proposed method of TiO_2 synthesis was superior to Degussa P-25, a commercial catalyst (most commonly used in experiments and in the field). Degussa P-25 is known to be a mixture of two polymorphs of TiO_2 (anatase and 25% rutile) in appropriate composition, limiting the electron-hole recombination and allowing the hole created to act as an effective oxidising agent.

Romanos and coworkers (2012) loaded the nanofiltration membrane with TiO_2 and used it effectively for filtration and photocatalytic treatment in a continuous reactor. Since the chemical vapour deposition (CVD) technique was used, TiO_2 could be deposited on both sides of the alumina nanofiltration membrane, which were irradiated with UV source. These proved to be highly efficient for the degradation of methyl orange and also inactivation of *E. coli* in a continuous-flow system.

6.2.11.2 Recent Studies on Disinfection

Robertson and coworkers (2012) effectively treated and removed the microorganisms and their chemical metabolites (cynotoxins), focussing on cynobacteria, using semiconductor photocatalyst. The paper covers 20 years of exclusive research highlighting basic processes and examples of photocatalyst-based microbial pollutant treatment and their chemical by-products.

Pigeot-Rémy and coworkers (2012) have tried to elucidate the difference in the mechanism and action of photochemistry and photocatalysis in obtaining bactericidal efficiency in water treatment. They observed that UVA, UVB, and UVC radiations affect the bacteria *E. coli* (model microorganism) differently and the mode of activation is different. Similarly, in photocatalysis, the OH^- radicals generated affect the outer microbial cell wall, increasing the permeability of other disinfectants and, in extreme cases, completing mineralisation of the lipid layers constituting the membrane cell wall, exposing the cytoplasm, which is vulnerable to external attack. Misstear and Gill (2012) have also elaborated the photolytic and photocatalytic action on bacteriophages (MS2, φX174, and PR772) and have concluded that the response of the phages is significantly different. The response was shown to depend on the type of phage, source of UV radiation (artificial or solar), and whether catalyst was present or not.

From the forgoing discussion one can conclude that the basic focus of research is in developing newer materials and newer reactor configurations. Immobilisation of the photocatalyst appears promising on a small to medium scale, as proposed correctly earlier, addressing smaller communities or temporary solutions in disaster-affected areas to provide safe and sanitised drinking water.

The conclusion and statement made by Hashimoto and coworkers (2007) is worth reproducing here as a concluding remark. After careful survey of 25 years of research and also taking into consideration the views of an industry, they concluded that TiO_2 photocatalysis could not be a practical technology for the following reasons.

It is fundamentally inadequate to utilise TiO_2 photocatalysis for either energy acquisition (H_2 generation) or the treatment of huge quantities of water or air, because light energy density is primarily low (only 3% of total solar radiation is in the UV range) and TiO_2 can only utilise the UV radiation of the solar spectrum.

Thus, the recent attempts in doping of TiO_2 with dye sensitisation appear to be the path forward. Also, cheaper solar concentrating technology, such as CPC or parabolic trough collector (PTC), in conjunction with a dye-sensitised TiO_2 or ruthenium-complexed TiO_2 appears to be a promising development. The combination and hybridisation of various processes such as filtration, photolysis, photocatalysis, and improving the utility of the entire solar spectrum, that is, UV visible part for OH^--based inactivation and IR part for thermal inactivation of the microbes, are suggested at this stage as the upcoming research area.

6.2.12 Reuse of Wastewater for Potable Use

It is well established that the current global water scenario is very bleak, and therefore other than water conservation, reuse of wastewater for various purposes such as agriculture, irrigation, and drinking is the need of the hour. This will definitely augment the existing drinking water supplies, and such type of water is commonly known as reclaimed water. The only major concern to this would be the potential health hazards given the plethora of chemical, physical, and biological contaminants in such wastewaters. However, it would be interesting to muse over the fact that the existing freshwater bodies are no longer clean owing to the dumping of wastes due to industrial activities. Therefore, the risks involved in reuse of wastewater may not exceed (except in terms of quantum or magnitude) those of the existing water supplies, and are definitely worth exploring, since they have great potential to contribute to any nation's available water resource. Another angle to this is the fact that wastewater reuse would mean that these are not released into water bodies, thereby reducing the stress already existing on the freshwater bodies. For several years, research on reuse of wastewater for uses other than drinking, such as for irrigation, construction of wetlands, and industrial uses like cooling tower water and even indirect potable use in the form of recharging groundwater, has been extensively carried out. The current research on its direct use as potable water is highlighted. This is a relatively new area and interesting articles are being published globally explaining the possible technologies for the same. In one such article published in the *Times of India, Bangalore*, the reuse of toilet waste into potable water is described. The investigation, funded by the Bill Gates Foundation, is all set to explore the use of a scaffold device holding a mixture of bacteria and tiny metal nanoparticles that are expected to react with the wastewater. The device is expected to extract useful hydrogen from the wastewater, which in turn would be converted to hydrazene or rocket fuel, and the remainder wastewater would be filtered to give clean water. Therefore, this study can be considered an inexpensive fuel-producing water-cleaning device. This investigation, led by a team of scientists from Manchester University, plans to have a prototype of the device ready to demonstrate by 2013. A major obstacle in the successful utilisation of wastewater reuse as drinking water is its public acceptance, especially if the waste that is being referred to is toilet flush water. However, people globally are becoming more aware of the water crisis and are slowly ready to accept these technologies. A recent article reports the success of San

Diego in implementing the toilet-to-tap concept after an initial struggle of over 12 years to make people accept the treated wastewater for the purpose of drinking (www.nytimes.com) (Barringer et al., 2012).

Questions

1. What are novel disinfection techniques? Emphasise the need for new technologies.
2. Describe plasma technology with respect to its fundamentals.
3. How can plasma be generated for water disinfection?
4. Explain the role of plasma methodology for wastewater treatment with specific examples.
5. Differentiate between thermal and nonthermal plasma methods for waste treatments.
6. Explain the role of plasma in potable water disinfection.
7. Define electrohydraulic discharge and its role in water disinfection with suitable diagrams.
8. Discuss any two recent potential economic techniques for water disinfection.
9. Explain what are online monitoring systems for disinfection and discuss some of the merits of such systems.
10. Define and explain the processes of electrocoagulation and electroflotation.
11. Discuss the role of nanotechnology in water disinfection, emphasising the recent developments in this area.
12. Briefly describe the very recent developments in solar water disinfection.
13. Describe ballast water treatment with respect to its need, and various techniques.
14. Explain the role of supercritical water oxidation in water disinfection with emphasis on the chemical reactions involved, operating range, and the engineering challenges faced.
15. Briefly explain the latest developments in membrane technology and its application for water disinfection.
16. Explain photocatalysis and its mechanism of disinfection.
17. Describe various photocatalytic reactors and their hybridisation.
18. Comment on the reuse of wastewater for potable purposes.

References

Alvarez-Ayuso, E., Garcia-Sanchez, A., and Querol, X. (2003, December). Purification of metal electroplating waste waters using zeolites. *Water Res.* 37(20):4855–4862.

Andra, S.S., Konstantinos, C., and Makris, J.P.S. (2011). Frequency of use controls chemical leaching from drinking-water containers subject to disinfection. *Water Res.* 45(20):6677–6687.

Anikeev, V.I., Yermakova, A., Mikenin, P.E., Belobrov, N.S., Piterkin, R.N., Prosvirnin, R.Sh., and Zvolsky, L.S. (n.d.). The present and future of the supercritical technologies in Russia; development of a mobile SCWO plant.

Bach, C., Dauchy, X., Chagnon, M.-C., and Etienne, S. (2012). Chemical compounds and toxicological assessments of drinking water stored in polyethyleneterephthalate (PET) bottles: A source of controversy reviewed. *Water Res.* 46(3):571–583.

Baker, R.W. (2000). *Membrane technology and applications.* New York: McGraw-Hill.

Barringer, F. (2012). As "yuck factor" subsides, treated wastewater flows from taps. http://www.nytimes.com/2012/02/10/science/earth/despite-yuck-factor-treated-wastewater-used-for-drinking.html?_r=1&pagewanted=all (accessed May 21, 2012).

Bej, A.K. (2003, May). Molecular based methods for the detection of microbial pathogens in the environment. *J. Microbiol. Meth.* 53(2):139–140.

Bhalchandra, A.P., Ranade, V.V., Chandrashekhar, A.A., Sawant, S.S., Ilangovan, D., Rajachandran, M., and Pilarisetty, V.K. (2010). U.S. Patent 7,815,810.

Blois, M.D. (2008). U.S. Patent 7,374,681.

Boldor, D., Balasubramanian, S., Purohit, S., and Rusch, K. (2008). Design and implementation of a continuous microwave heating system for ballast water treatment. *Environ. Sci. Technol.* 42(11):4121–4127.

Browning, W.J. (2001). U.S. Patent 6,171,508.

Chang, J.S. (2001). Recent development of plasma pollution control: Critical review. *Sci. Technol. Adv. Mater.* 2:571–576.

Chase, C., Reilly, C., and Pederson, J. (2007). *Marine bioinvasions fact sheet: Ballast water treatment options.* Cambridge, MA: MIT Sea Grant Center for Coast Resources, 2001.

Chu, H., Dong, B., Zhang, Y., Zhou, X., and Yu, Z. (2012). Pollutant removal mechanisms in a bio-diatomite dynamic membrane reactor for micro-polluted surface water purification. *Desalination* 293(1):38–45.

Cronholm, L.S., MacCammon, J.R., and Kruse, C.W. (1976). Enteric viruses survival in package plants and the upgrading of the small treatment plants using ozone. Research Report 98. Lexington: Water Resources Institute, University of Kentucky.

Diallo, M.S. (2004). Water treatment by dendrimer enhanced filtration. U.S. Patent pending.

Diallo, M.S., Christie, S., Swaminathan, P., Johnson Jr., J.H., and Goddard III, W.A. (2005). Dendrimer enhanced ultrafiltration. Recovery of Cu(II) from aqueous solutions using PAMAM dendrimers with ethylene diamine core and terminal NH2 groups. *Environ. Sci. Technol.* 39:1366–1377.

Dors, M. (2011). Plasma for water treatment. http://www.plastep.eu/fileadmin/dateien/events/2011/110725_summer_school/plasma_water_treatment.pdf (accessed May 16, 2012).

Fugetsu B., Satoh, S., Shiba, T., Mizutani, T., Lin, Y.B., and Terui, N. (2004). Caged multiwalled carbon nanotubes as the adsorbents for affinity-based elimination of ionic dyes. *Environ. Sci. Technol.* 38:6890–6896.

Fisher, M.B., Iriarte, M., and Nelson, K.L. (2012). Solar water disinfection (SODIS) of *Escherichia coli*, *Enterococcus* spp., and MS2 coliphage: Effects of additives and alternative container materials. *Water Res.* 46(6):1745–1754.

Frank, S.N., and Bard, A.J. (1977). Heterogeneous photocatalytic oxidation of cyanide ion in aqueous solutions at titanium dioxide powder. *J. American Chem. Soc.* 99:303–304.

Gaot, W., Majumder, M., Alemany, L.B., Narayanan, T.N., Ibarra, M.A., Pradhan, B.K., and Ajayan, P.M. (2011). Engineered graphite oxide materials for application in water purification. *ACS Appl. Mater. Interfaces* 3(6):1821–1826.

Guddeti, R.R., Knight, R., and Grossmann, E.D. (2000). Depolymerization of polyethylene using induction-coupled plasma technology. *Plasma Chem. Plasma Proc.* 20:37–64.

Goldstein, B.D., and McDonagh, E.M. (1975). Effect of ozone on cell membrane protein fluorescence. I. *In vitro* studies utilizing the red cell membrane. *Environ. Res.* 9(2):179–186.

Gomez, E., Amutha Raniab, D., Cheesemanb, C.R., Deeganc, D., Wisec, M., and Boccaccini, A.R. (2009). Thermal plasma technology for the treatment of wastes: A critical review. *J. Hazard. Mater.* 161:614–626.

Gregg, M.D., and Gustaaf, M.H. (2007). Efficacy of three commercially available ballast water biocides against vegetative microalgae, dinoflagellate cysts and bacteria. *Harmful Algae* 6:567–584.

Guo, H., and Hu, J. (2012). Effect of hybrid coagulation—Membrane filtration on downstream UV disinfection. *Desalination* 290(30):115–124.

Hashimoto, K., Irie, H., and Fujishima, A. (2007). TiO_2 photocatalysis: A historical overview and future prospects. *AAPPS Bull.* 17(6):12–28.

http://www.turbosynthesis.com/summitresearch/sumscw1.htm (accessed May 12, 2012).

Hu, J., How Yong, N.G., and Say Leong, O.N.G. (2012). An overview of current membrane technologies and possible future developments for the removal of emerging contaminants in water reclamation. COVER STORY: *Membrane Technology: Future of Water Reclamation*, http://www.innovationmagazine.com/innovation/volumes/v8n1/coverstory2.shtml (accessed September 29, 2012).

Huang, H., and Tang, L. (2007, April). Treatment of organic waste using thermal plasma pyrolysis technology. *J. Environ. Conserv. Mgmt.* 48(4):1331–1332.

Jin, X., Zhang, H., Wang, X., and Zhou, M. (2012). An improved multi-anode contact glow discharge electrolysis reactor for dye discoloration. *Electrochim. Acta* 59(1):474–478.

Kazumi, J. (2007). *Ballast water treatment technologies and their application*. Transportation Research Board Special Report 291.

Khaydarov, R., Khaydarov, R., and Yuldashev, B. (2007). Experience of using energy-effective water disinfection devices. http://213.55.83.52/ebooks//Construction%20Management/16742.pdf (accessed May 20, 2012).

Kogelschatz, U. (2000). Fundamentals and applications of dielectric-barrier discharges. http://www.coronalab.net/wxzl/plasma-16.pdf (accessed May 10, 2012).

Kuzirian, A.M., Eleanor, C., Terry, S., Bechtel, D.L., and Patrick, J.L. (2001). Hydrogen peroxide: An effective treatment for ballast water. *Biol. Bull.* 201:297–299.

Lerf, A., He, H., Forster, M., and Klinowski, J. (1998). Structure of graphite oxide revisited *J. Phys. Chem. B* 102(23):4477–4482.

Li, Q.L, Yuan, D.X., and Lin, Q.M. (2004). Evaluation of multiwalled carbon nanotubes as an adsorbent for trapping volatile organic compounds from environmental samples. *J. Chromatogr.* 1026:283–288.

Liu, L., Liu, Z., Bai, H., and Sun, D.D. (2012). Concurrent filtration and solar photocatalytic disinfection/degradation using high-performance Ag/TiO2 nanofiber membrane. *Water Res.* 46(4):1101–1112.

Lloyd's Register. (2007). *Ballast water treatment technology.*

Lloyd's Register. (2011). *Ballast water treatment technology.*

Locke, B.R., Sato, M., Sunka, P., Hoffman, M.R., and Chang, J.S. (2005). Electrohydraulic discharge and non-thermal plasma for water treatment. *Ind. Eng. Chem. Res.* 45:882–905.

Loo, S., Fane, A.G., Krantz, W.B., and Lim, T.-T. (2012). Emergency water supply: A review of potential technologies and selection criteria. *Water Res.*, 46(10):3125–3151.

Malik, M.A., Ghaffar, A., and Malik, S.A. (2001). Water purification by electrical discharges. *Plasma Sources Sci. Technol.* 10:82–91.

Mangun, C.L., Yue, Z.R., Economy, J., Maloney, S., Kemme, P., and Cropek, D. (2001). Adsorption of organic contaminants from water using tailored ACFs carbon. *Chem. Mater.* 13:2356–2360.

Matteson, M.J., Dobson, R.L., Glenn, R.W.J., Kukunoor, N.S., Waits, W.H.I. and Clayfield, E.J. (1995). Electrocoagulation and separation of aqueous suspensions of ultrafine particles. *Colloids Surf., A* 1041:101–109.

Mbonimpa, E.G., Vadheim, B., and Blatchley III, E.R. (2012). Continuous-flow solar UVB disinfection reactor for drinking water. *Water Res.* 46(7):2344–2354.

McNulty, P.D. (2003). U.S. Patent 0205136 A1.

Misstear, D.B., and Gill, L.W. (2012). The inactivation of phages MS2, ΦX174 and PR772 using UV and solar photocatalysis. *J. Photochem. Photobiol. B Biol.* 107(6):1–8.

Mohai, I., and Szepvolgyi, J. (2005). Treatment of particulate metallurgical wastes in thermal plasmas. *Chem. Eng. Process.* 44:225–229.

Moreno, N., Querol, X., and Ayora, C. (2001). Utilization of zeolite synthesized from coal fly ash for the purification of acid. *Environ. Sci. Technol.* 35:3526–3534.

Nanayakkara, K.G.N., Yu-Ming, Z., Khorshed, A., Shuaiwen, Z., and Chen, P.J. (2011). Electrochemical disinfection for ballast water management: Technology development and risk assessment. *Marine Pollut. Bull.* 63(5–12):119–123.

National Research Council. (1996). *Controlling introductions of nonindigenous species.* Washington, DC: National Academies Press.

Nielsen, B.C. (2006). Control of ballast water organisms with a seawater electrochlorination and filtration system. M.S. thesis, University of Washington.

Oemcke, D.J., and Leeuwen, J.V. (2005). Ozonation of the marine dinoflagellate alga. *Water Res.* 39:5119–5125.

O'Regan, B., and Grätzel, M. (1991). A low-cost, high-efficiency solar cell based on dye-sensitized colloidal TiO_2 films. *Nature* 353:737–740.

Ottaviani, M.F., Favuzza, P., Bigazzi, M., Turro, N.J., Jockusch, S., and Tomalia, D.A. (2000). TEM and EPR investigation of the competitive binding of uranyl ions to starburst dendrimers and liposomes. Potential use of dendrimers as uranyl ion sponges. *Langmuir* 16:7368–7372.

Papageorgiou, S.K., Katsaros, F.K., Favvas, E.P., Romanos, G.Em., Athanasekou, C.P., Beltsios, K.G., Tzialla, O.I., and Falaras, P. (2012). Alginate fibers as photocatalyst immobilizing agents applied in hybrid photocatalytic/ultrafiltration water treatment processes. *Water Res.* 46(6):1858–1872.

Peng, X.J., Li, Y.H., Luan, Z.K., Di, Z.C., Wang, H.Y., Tian, B.H., and Jia, Z.P. (2003). Adsorption of 1,2-dichlorobenzene from water to carbon nanotubes. *Chem. Phys. Lett.* 376:154–158.

Perrut, M. (2012). Sterilization and virus inactivation by supercritical fluids [a review]. *J. Supercritical Fluids* 66:359–371.

Peter-Varbanets, M., Hammes, F., Vital, M., and Pronk, W. (2010). Stabilization of flux during dead-end ultralow pressure ultrafiltration. *Water Res.* 44(12):3607–3616.

Peter-Varbanets, M., Gujer, W., and Pronk, W. (2012). Intermittent operation of ultra-low-pressure ultrafiltration for decentralized drinking water treatment. *Water Res.* 46(10):3272–3282.

Pigeot-Rémy, S., Simonet, F., Atlan, D., Lazzaroni, J.C., and Guillard, C. (2012). Bactericidal efficiency and mode of action: A comparative study of photochemistry and photocatalysis. *Water Res.* 46(10):3208–3218.

Razi, F., Sawada, I., Ohmukai, Y., Maruyama, T., and Matsuyama, H. (2012). The improvement of antibiofouling efficiency of polyethersulfone membrane by functionalization with zwitterionic monomers. *J. Membr. Sci.* 401–402:292–299.

Ricordel, C., Darchen, A., and Hadjiev, D. (2010). Electrocoagulation–electroflotation as a surface water treatment for industrial uses. *Separation Purification Technol.* 74(3):342–347.

Riesser, V.W., et al. (1976). Possible mechanisms of poliovirus inactivation by ozone. In E.G. Fochtman, R.G. Rice, and M.E. Browning (eds.), *Forum on ozone disinfection*. Syracuse, NY: International Ozone Institute, pp. 186–192.

Robertson, P.K.J., Robertson, J.M.C., and Bahnemann, D.W. (2012). Removal of microorganisms and their chemical metabolites from water using semiconductor photocatalysis. *J. Hazard. Mater.* 211–212:161–171.

Romanos, G.Em., Athanasekou, C.P., Katsaros, F.K., Kanellopoulos, N.K., Dionysiou, D.D., Likodimos, V., and Falaras, P. (2012). Double-side active TiO_2-modified nanofiltration membranes incontinuous flow photocatalytic reactors for effective water purification. *J. Hazard. Mater.* 211–212:304–316.

Ryu, H.S., Ha, Y.C., Seo, D.H., and Kim, D.E. (2010). U.S. Patent 2010/0126940.

Sanaeepur, H., Hosseinkhani, O., Kargari, A., Amooghin, A.E., and Raisi, A. (2012). Mathematical modeling of a time-dependent extractive membrane bioreactor for denitrification of drinking water. *Desalination* 289(15):58–65.

Sano, L.L., Moll, R.A., Krueger, A.M., and Landrum, P.F. (2003). Assessing the potential efficacy of glutaraldehyde for biocide treatment of un-ballasted transoceanic vessels. *J. Great Lakes Res.* 29(4):545–557.

Scott, D.B.M.N., and Lesher, E.C. 1963. Effect of ozone on survival and permeability of *Escherichia coli*. *J. Bacteriol.* 85:567–576.

Sharma, V.K., Theodoros, M., Triantis, M.G., Antoniou, X.H., Pelaez, M., Han, C., Song, W., O'Shea, K.E., de la Cruz, A.A., Kaloudis, T., Hiskia, A., and Dionysios, D. (2012). Destruction of microcystins by conventional and advanced oxidation processes: A review. *Separation Purification Technol.* 91(3):3–17.

Son, W.K., Youk, J.H., and Park, W.H. (2006). Antimicrobial cellulose acetate nanofibers containing silver nanoparticles. *Carbohydr. Polym.* 65:430–434.

Sontakke, S., Mohan, C., Modak, J., and Madras, G. (2012). Visible light photocatalytic inactivation of *Escherichia coli* with combustion synthesized TiO2. *Chem. Eng. J.* 189–190(1):101–107.

Steeper, R.R. (1996, January). Methane and methanol oxidation in supercritical water: Chemical kinetics and hydrothermal flame studies. Sandia National Laboratory Report SAN96-8208-UC-1409.

Stoquart, C., Servais, P., Bérubé, P.R., and Barbeau, B. (2012). Hybrid membrane processes using activated carbon treatment for drinking water: A review. *J. Membr. Sci.* 411/412:1–12.

Subrahmanyama, Ch., Renkenb, A., and Kiwi-Minsker, L. (2010). Catalytic non-thermal plasma reactor for abatement of toluene. *Chem. Eng. J.* 160:677–682.

Tamburri, M.N., Wasson, K., and Matsuda, M. (2002). Ballast water deoxygenation can prevent aquatic introductions. *Biol. Conserv.* 103:331–341.

Tang, Z., Butkus, M.A., and Xie, Y.F. (2005). Crumb rubber filtration: A potential technology for ballast water treatment. *Marine Environ. Res.* 61:410–423.

Van Rijn, C.J.M. (2004). *Nano and microengineered membrane technology.* Amsterdam: Elsevier.

Wakao, Y., Takuro, T., and Takashi, M. (2006). U.S. Patent 2006/0289364 A1.

Wang, L.K., Chen, J.P., Hung, Y.-T., and Shammas, N.K. (2011). *Membrane and desalination technologies series: Handbook of environmental engineering*, Vol. 13, 1st ed. New York: Springer.

Wright, D.A., Gensemer, R.W., Mitchelmore, C.L., and Stubblefield, W.A. (2010). Shipboard trials of an ozone-based ballast water treatment system. *Marine Pollut. Bull.* 60:1571–1583.

Zeman, L.J., and Zydney, A.L. (1996). *Microfiltration and ultrafiltration.* New York: Marcel Dekker.

Zuo, Q., Chen, X., Li, W., and Chen, G. (2008). Combined electrocoagulation and electroflotation for removal of fluoride from drinking water. *J. Hazard. Mater.* 159(2–3):452–457.

7

Summary

Potable water, the elixir of life, will continue to be a vital necessity for humankind as long as life exists. This also implies that not only access to drinking water but also its safety is imperative. The quest for safe drinking water has been the priority for humans over centuries and will continue to dominate the world forever. This can be attributed to the overwhelming rise in the global population, depletion of natural resources, and ill-managed water systems, especially the lack of effective disinfection and treatment methods, specifically in rural areas. We, as individuals as well as part of a society, need to create awareness about the existing scenario and also endeavour to motivate every human to take small baby steps toward obtaining clean and safe drinking water, which is our birthright! Therefore, an attitudinal change is the need of the hour; that is, try doing things on your own and assist the authorities in meeting your potable water requirements.

This book described a variety of disinfection techniques for potable water, such as chemical, physical, and hybrid methods, including traditional processes such as treatment with chlorine and sand filtration, on one hand, and novel emerging techniques like cavitation, on the other. All the methods described for inactivation of microbes in this book had their merits and demerits, and no individual or hybrid process can be accepted as a universal remedy or solution for obtaining safe potable water. This may be ascribed to a plethora of factors that have been described in detail previously and can be emphasised here again. Some of the concerning factors are the natural source of water, its physical, chemical, and biological quality, environmental factors like pH and temperature, end use, and most importantly, the economics of the proposed disinfection process.

7.1 Global Water Challenges

It has been harped on over and over again that water-related issues such as water shortage, improper water quality, and uneven special distribution are likely to augment pressures globally in the next decade. It has also been predicted that countries are likely to wage war against each other over water. However, this is only one side of the coin. The other side is the possibility of water-sharing agreements among nations rather than conflicts. Nevertheless, the areas that are likely to be most affected are North Africa, the Middle East, and South Asia. Globally, leaders of various nations are already addressing this issue. In the recent World Water Day celebrations on March 22, 2012, U.S. Secretary of State Hillary Clinton, clearly emphasised U.S. commitments to world water challenges. "We believe this will help map our route to a more water-secure world," Clinton said, "a world where no one dies from water-related diseases, where water does not impede social or economic development, and where no war is ever fought over water" (http://iipdigital. usembassy.gov). Social and economic development is hindered when millions of women and girls have to travel for miles every day in order to fetch water for bare necessities. Clinton pointed out the amazing results that could be attained by partnership of bodies with different core competencies, such as identifying and collecting a database of available water sources, water supply management, and technology for treatment of water, as well as minimising disruption of the environment and conservation of natural resources. Therefore, it is really a collective endeavour to deliver clean and safe drinking water to humankind with each and every individual or a body contributing to the final objective. It is very concerning to note that the U.S. National Intelligence Council, in its global water assessment report, has stated that several countries most important to the United States will be facing drastic water problems that may lead to some instability within different countries, diverting their attention from other important U.S. policy objectives. It has also been envisaged that freshwater resources will not suffice the population in the future. Also, water issues will have a repercussion on the country's ability to produce food and generate electricity, thereby posing a risk to the food markets and hampering economic growth. Moreover, for achieving global health improvement, clean and safe water is imperative. Besides, opportunities in education and employment for women cannot be achieved when they have to spend several hours each day in the quest for water.

7.2 Current Activities to Solve Water-Related Issues

Several important governmental bodies and nongovernmental organisations (NGOs) have come together in an attempt to bring about positive changes to the pressing water-related problems. Around 30 activities have been identified where various U.S. agencies are expected to work more closely with the World Bank and with each other. The U.S. Agency for International Development (USAID) and the National Aeronautics and Space Administration (NASA) are working in synergy to analyse water security and other water-related issues in some countries. A very remarkable initiative by USAID is the launching of Water, Sanitation and Hygiene (WASH)—the wash for life partnership with the Gates Foundation. This initiative will help in delivering these services to some of the poorest and most rural regions of the world. Similar programmes by USAID are helping to prevent the spread of many waterborne diseases. It is heartening to note that the objective of the Millennium Development Goal to cut in half the proportion of people living without access to safe drinking water was reached almost four years ahead of schedule (in 2010) as announced by the United Nations recently. This is very important and remarkable progress made, although several milestones still exist to be reached toward the final goal. Therefore, clean water and access to water are of paramount importance to future peace, security, and prosperity of humankind worldwide.

Among the various water challenges existing globally, reuse of water is an important aspect that has to be addressed. This is of utmost significance since global water resources are getting depleted and effective water reuse will be very critical to the sustainable functioning of industry, agriculture, and people around the world. The U.S. Environmental Protection Agency (USEPA) has published guidelines (developed by CDM Smith in 1980 and updated in 1992 and 2004) for water reuse, which throws light on some important guidance on water reusage to water and wastewater utilities, regulatory agencies, and industries worldwide. In fact, the USEPA, USAID, and CDM Smith have already started development of the next EPA guidelines for water reuse document, which will address some of the most pressing issues. The 2012 updated report will provide information on a spectrum of water reuse practices, from simple low-technology solutions involving stage-wise water usage (discharge of one activity as an input to others) to advanced treatment methods. Areas not covered in the previous version of the report, such as reuse in integrated water resource planning and management, energy use, and grey water reuse systems, as well as direct reuse for potable options, will be highlighted. It is really important to understand that agriculture reuse is going to play a bigger role than it has in the past, in water reuse supply and demand, and the current report will emphasise this aspect, especially use of reclaimed water for crops. It is also proposed to create a Web site (waterreuseguidelines.org) to provide access to the latest information. This report will be very important, especially for developing countries as they can gain access to better management practices, especially in the area of wastewater reuse with minimal human health risks. The primary goal of the report is to provide a variety of options that countries globally can employ to make reused water a safe and reliable water source. These updates are scheduled to be released at the annual Water Environment Federation Technical Exhibition and Conference in October 2012 (www.cdmsmith.com).

A central issue to be addressed among the major global water challenges is to focus on the rural areas to provide them with basic sanitation and safe drinking water. It is very concerning to note that poor people in all parts of the world are lacking in these basic necessities. The report Progress on Drinking Water and Sanitation 2012 states that around 89% of the world's urban population used better drinking water sources. However, it is of paramount importance to understand that the report also mentions that the rural areas are still without access to safe drinking water compared to the urban areas. Statistics from UNICEF indicates that the rural-urban divide for drinking water is very pronounced. In 2010 there were 783 million people worldwide without access to sources of safe drinking water, out of which 653 million were from rural regions. Alarmingly, it is women and young girls who are shouldering the water burden in these rural households. They have to walk long distances to fetch water every day, and a UNICEF survey conducted in 25 countries in sub-Saharan Africa found that in 71% of all households without drinking water, women and girls were responsible for water collection, which amounted to one or more trips to the site. It is very distressing to note that women, men, and children spend a combined total of at least 16, 6, and 4 million man-hours each day respectively collecting drinking water (www. unicef.org). With the gamut of challenges and the steps and measures being undertaken worldwide, it will be worthwhile to also know where the global water markets are heading.

7.3 World Water Markets

A study reported at hkc22.com by Helmut Kaiser Consultancy, which has been a leading consulting company in the area of water industry worldwide for more than two decades, demonstrates high growth rates in the global water markets. The competition is growing among companies worldwide and is expected to grow tougher. Statistics represent 2000 companies globally that lead the world markets for water. Asia and the Middle East show the highest growth rates, and China is leading the race in the market. Significant water quality technologies are from Europe and the United States. The major focus worldwide is on water quality and clean water, followed by bottled water and water for the food industry. It is very interesting to note that disinfection, homeland security, and desalination are considered the most profitable markets (http://www.hkc22.com/water.html). Interestingly, the cost of water in the future is predicted to depend on the disinfection requirement and the technologies applied to reach the objective. The market for water disinfection is stated to be expanding between 8 and 25% (according to October 2011 data) based on the region and country. Most recent technological advancements in this arena are reported as nanotechnology and molecular technologies, and these technologies are expected to change the water treatment markets in the next decade.

Thus, to sum up, the global scenario is set with several challenges for water. Most importantly, water is becoming scarce and judicious use of existing water has become imperative. Reuse of wastewater, including toilet flush water (discussed in Chapter 6), is currently being researched. It will be thus appropriate to state that today's potable water was yesterday's waste!

Disinfection of water is very critical, as newer contaminants such as pesticides, pharmaceuticals, and personal care products are being added to the water bodies. However, this requires thorough study of the contaminants before a treatment can be envisaged, and also reworking the water standards and the permissible limits of the emerging contaminants. Keeping with these pressures, research on water disinfection is speeding up globally with relevant government bodies funding such projects. However, it appears that the outcomes of many of these research endeavours, in spite of being effective, are not being translated into large-scale treatment. This aspect has to be addressed, keeping in mind the criticality of the current situation. Collectively handling the current scenario by conserving water with activities like rainwater harvesting, reusing wastewater, and employing effective yet economical large-scale disinfection methods, with the main focus on rural areas, will definitely lead to a positive bright future with a very high global water quotient.

References

Clinton boosts U.S. commitment to world water challenges. (2012). http://iipdigital. usembassy.gov/st/english/article/2012/03/201203222612.html (assessed May 7, 2012).

Drinking water and waste water 2015 worldwide by regions (Bn. Euro). (2012).

http://www.hkc22.com/water.html (assessed May 7, 2012).

Progress on drinking water and sanitation. (2012). http://www.unicef.org/media/files/JMPreport2012.pdf (assessed May 2, 2012).

Seeing clearly: Water quality redefined. (2012). http://cdmsmith.com/en-AP/insights/features/seeing-clearly-water-quality-redefined.aspx (assessed April 29, 2012).

8

Problems

1. Considering UV disinfection of a water sample contaminated with coliforms, what will be the effluent viable coliform density per 100 ml (N), for the following data:

UV dose (D) = 150 m.W.s/cm^2
Empirical coefficient (n) = 10
Empirical water quality factor (f) = 2.4

Sample calculation:

$N = f \cdot Dn$ (Tchobanoglous, 1997)
$N = 2.4 \times 150 \times 10$
$N = 3600/100$ ml

2. Water contaminated with faecal coliforms is subjected to disinfection in a UV module. If the effluent coliform density was 2133/100 ml, what will be the UV dose required for complete disinfection?

Empirical coefficient (n) = 15
Empirical water quality factor (f) = 2.9

(Hint: Use the equation suggested by Tchobanoglous (1997): $N = f \cdot Dn$.)

3. Water contaminated with pathogens is being treated with UV to make it potable. If the load of coliforms is 3027/100 ml, what will be the UV dose required to disinfect the water sample to make it?

Empirical coefficient (n) = 14
Empirical water quality factor (f) = 1.3

(Hint: Use the equation suggested by Tchobanoglous.)

4. Contaminated water with coliform density of 2030/100 ml was disinfected with a UV dose of 125 m.W.s/cm². What will be the empirical water quality factor?

   Empirical coefficient (n) = 6

   (Hint: Use the equation suggested by Tchobanoglous.)

5. Calculate the percentage UV transmittance (% UVT) at a wavelength of 288 nm and a path length of 1 cm for an absorbance value of 1.0.

   Sample calculation:

   UVT can be calculated by relating it to UV absorbance using the following equation:

   % UVT = 100×10^{-A}

   where A is absorbance, UVT = UV transmittance at a specified wavelength (e.g., 254 nm) and path length (e.g., 1 cm), A = UV absorbance at a specified wavelength and path length (unitless), and

   % UVT = 100×10^{-A}
   $\quad\quad\quad = 100 \times 10^{-1}$
   $\quad\quad\quad = 10$

6. Calculate the percent UVT at a wavelength of 285 nm and path length of 1.5 cm, given that the absorbance is 0.5. (Hint: % UVT = 100×10^{-A}).

7. Calculate the UVT if the intensity of light transmitted through water is 266 mW/cm² and the intensity of the light incident on the water sample is 302 mW/cm².

   Sample calculation:

   UVT is the percentage of light passing through material over a specified distance. The UVT can be calculated using Beer's law:

   UVT = $100 * I / I_o$

   where UVT = UV transmittance at a specified wavelength (e.g., 254 nm) and path length (e.g., 1 cm), I = intensity of light transmitted through the sample (milliwatt per centimetre squared (mW/cm²)), I_o = intensity of light incident on the sample (mW/cm²), and

   UVT = $100 * I / I_o$
   $\quad\quad = 100 \times 266 / 302$
   $\quad\quad = 88.07$

8. Calculate the UV transmittance for a UV disinfecting module if the intensity of light transmitted through water is 280 mW/cm² and the intensity of the light incident on the water sample is 325 mW/cm².

   (Hint: UVT = 100*I/Io.)

9. If water having a streptococcal load of 2987 CFU/ml was treated with UV irradiation and the concentration of the microbes was reduced to 546 CFU/ml, comment on the microbial response.

   Sample calculation:

   Microbial response is a measure of the sensitivity of the microorganism to UV light and is unique to each microorganism. UV dose-response is determined by irradiating water samples containing the microorganism with various UV doses and measuring the concentration of infectious microorganisms before and after exposure. The microbial response is calculated using the following equation:

   Log inactivation = $\log_{10} N/N_0$

   where N_0 = concentration of infectious microorganisms before exposure to UV light, N = concentration of infectious microorganisms after exposure to UV light, and

   Log inactivation of streptococci = $\log_{10} 546/2987$

   $$= -0.738$$

10. If water having an *E. coli* load of 3290 CFU/ml was treated with UV irradiation and the concentration of the microbes reduced to 1200 CFU/ml, comment on the microbial response. A similar experiment was conducted using water with coliform density of 4362/100 ml, and after UV treatment the coliforms were inactivated reducing their concentration to 200/100 ml. Comment on the coliform response to UV irradiation and also mention which microorganism is susceptible to UV treatment.

    (Hint: Log inactivation = $\log_{10} N/N_0$.)

11. What will be the thickness of membrane required to treat water if the flux is 27 L/m².h and pressure drop is 60 N/m² and the osmotic pressure difference is negligible, and consider the specific permeability value for water as 1.9×10^{-7} g/cm².s.

Sample calculation: Consider the water treatment process using membranes. The flux equation for the process is

$$N_w = P_w \frac{\Delta P - \Delta \pi}{L}$$

where N_w = flux, P_w = specific membrane permeability, $g/cm^2.s$, ΔP = pressure across membrane, $\Delta \pi$ = osmotic pressure difference, and L = thickness of membrane. The specific permeability values for water are in the range of 1.6 to $3.8 \times 10^{-7} kg.m^2.h$.

Substituting in the above equation we get,

$N_w = 27 \, l/m^2.h$
$P_w = 1.9 \times 10^{-7} \, kg/m^2 h$
$\Delta P = 60 \, N/m^2$
$27 = 1.9 \times 10^{-7} \times 60/l$
$L = 4.22 \times 10^{-7} \, m$

12. If the thickness of membrane required to treat water was 0.9 cm, calculate the flux applied for a pressure drop of 43 N/m^2 and an osmotic pressure difference of 10 N/m^2. Consider the specific permeability value for water as 2.2×10^{-7} $g/cm^2.s$.

    (Hint: $N_w = P_w (\Delta P - \Delta \pi)/l$.)

13. If a 7000 gal water tank has to be disinfected with at least a 200 ppm solution of chlorine in it, how much 6.00% sodium hypochlorite (laundry bleach) would be needed?

    Sample calculation:

    Volume of hypochlorite = volume of water × (required residual in ppm/1,000,000 × hypochlorite %)

    (7,000 gal × 200 ppm)/(1,000,000 × 0.06) = 23.33 gal

14. For the same capacity of water tank and ppm level, how much 12% sodium hypochlorite solution would be needed?

15. If 6000 gal of water has to be disinfected by using calcium hypochlorite, how many pounds of 60% calcium hypochlorite solution is needed to achieve 200 ppm?

Sample calculation:

Weight of calcium hypochlorite (lb)
= gal of water × 8.33 lb/gal × (required residual in ppm/1,000,000 × hypochlorite %) (http://www.maine.gov)
(8.33 lb/gal × 6000 gal × 200 ppm)/(1,000,000 × 0.06) = 166.6 lb

16. If 8000 gal of water has to be disinfected by using calcium hypochlorite, how many pounds of 65% calcium hypochlorite is needed to achieve 100 ppm?

17. Consider water disinfection using ozone. If the concentration of ozone used was 0.1 kg/m³, assuming $n = 1$ and the CT value constant for ozone disinfection is 2.9×10^{-3} kg.min, what will be the contact time needed to achieve bacterial inactivation?

   Sample calculation:

   $k = C_n \times t$ (Viessman and Hammer, 1993)
   $2.9 \times 10^{-3} = 0.1 \times t$
   $t = 290$ min

18. Consider water disinfection using chlorine dioxide. If the concentration of chlorine dioxide used was 0.25 kg/m³, assuming $n = 1$ and the contact time needed for disinfection as 10 min, what will be the CT value constant for disinfection?

   (Hint: $k = C^n \times t$.)

19. Calculate the transmembrane pressure for a water disinfection operation through a typical ultrafiltration module if the pressure at the inlet of the membrane module is 230 psi and the pressure at the outlet of the membrane is 189 psi, given that the permeate pressure is 200 psi.

   Sample calculation:
   According to Tutujian (1985):

   $Ptm = (Pi + Po) - Pp,$

   where
   Ptm = transmembrane pressure (psi),
   Pi = pressure at the inlet of the membrane module (psi),
   Po = pressure at the outlet of the membrane module (psi), and
   Pp = permeate pressure (psi).

$Ptm = (230 + 189) - 200$ psi

$\quad = 219$ psi

20. Calculate the permeate pressure for a water disinfection operation through a typical ultrafiltration module if the pressure at the inlet of the membrane module is 330 psi and the pressure at the outlet of the membrane is 200 psi, given that the transmembrane pressure is 248 psi.

    (Hint: $Ptm = (Pi + Po) - Pp$.)

21. What will be the permeate flux at a temperature of 20°C if the permeate flow through a nanofiltration module is 20 l/h and the membrane surface area is 40 m²? The temperature of filtration is 35°C.

    Sample calculation:

    The equation proposed by Jacangelo et al. (1994) is:

    $J20 = Qp$ e-0.0239 $(T - 20)/S$

    where
    $J20$ = permeate flux corrected for 20°C (l/m²h),
    Qp = permeate flow (l/h),
    T = permeate test temperature (°C), and
    S = membrane surface area (m²).

    $J20 = 20$ e-0.0239 $(35 - 20)/40$
    $J20 = 62.29$ l/m²h

22. What will be the permeate flux at a temperature of 20°C if the permeate flow through a nanofiltration module is 80 l/h and the membrane surface area is 50 m²? The temperature of filtration is 40°C.

    (Hint: $J20 = Qp$ e-0.0239 $(T - 20)/S$.)

23. What will be the specific permeability flux for water passing through an ultrafiltration module if the permeate flux corrected at 20°C is 212 l/m²h and the transmembrane pressure is 30 kPa?

    Sample calculation:

    Specific flux (e.g., permeability) is defined as the ratio of permeate flux to transmembrane pressure (TMP) according to the following equation:

    $Jsp = J20/Ptm$

where
Jsp = specific flux (l/m²h kPa) and
Ptm = transmembrane pressure (kPa).

$Jsp = J20/Ptm$
$Jsp = 212/30$
$Jsp = 7 \, l/m²h \, kPa$

24. What will be the specific permeability flux for water passing through an ultrafiltration module if the permeate flux corrected at 20°C is 435 l/m²h and the transmembrane pressure is 70 kPa?

   (Hint: $Jsp = J20/Ptm$.)

25. Calculate the percent recovery for a microfiltration membrane if the permeate flow and the feed water flow rate are 50 and 70 l/h, respectively.
   Sample calculation:
   Recovery is defined as the ratio of permeate flow to feed water flow rate according to the following equation:

   $\% \, R = QP/QF$

   Low-pressure membranes (microfiltration [MF] and ultrafiltration [UF]) typically operate within the range of 85 to 97% recovery:

   $\% \, R = 50/70$
   $\% \, R = 0.71$

26. Calculate the permeate flow rate if the percent recovery for a microfiltration membrane and the feed water flow rate are 85% and 90 l/h, respectively.

   (Hint: $\% \, R = QP/QF$.)

27. One hundred millilitres of contaminated water containing an initial microbial load of 6980 CFU/ml was subjected to ultrasonication using a probe type sonicator. After 15 min of treatment the microbial load was found to be 3500 CFU/ml. Assuming a first-order dependence, estimate the overall rate of disinfection by ultrasonic horn.

   Sample calculation:

Initial CFU/ml = 6980 CFU/ml

CFU/ml at the end of 15 min of treatment = 3500 CFU/ml

Rate = no. of CFU killed/s, $-\dfrac{dc}{dt} = kC$ (assuming first-order dependence)

$$\therefore -\int_{C_1}^{C_2} \dfrac{dc}{C} = k\int_{0}^{t} dt$$

$$\therefore \ln\dfrac{C_1}{C_2} = kt$$

$$\therefore k = \left(\ln\dfrac{C_1}{C_2} \right)/t$$

$k = (\ln 6980/3500)/900$

$\quad = 7.67 \times 10^{-4}\,\text{s}^{-1}$

$C = C_1 + C_2/2$

$C = 6980 + 3500/2$

$C = 5240\ \text{CFU/ml}$

Vol. = 100 ml

Thus,

Rate of disinfection $= -\dfrac{dc}{dt} = kC, \dfrac{\text{CFU}}{\text{ml.s}}$

Rate of disinfection $= 7.67 \times 10^{-4} \times 5240$

$\qquad\qquad\qquad\quad = 4.02\ \text{CFU/ml.s}$

Overall rate of disinfection $= -\dfrac{dc}{dt} = k \times C \times \text{vol}, \dfrac{\text{CFU}}{\text{s}}$

$\qquad\qquad\qquad = 7.67 \times 10^{-4} \times 5240 \times 100$

$\qquad\qquad\qquad = 402\ \text{CFU/s}$

28. One litre of raw water containing an initial microbial load of 9800 CFU/ml was subjected to high-speed homoginisation. After 30 min of treatment the microbial load was found to be 4500 CFU/ml. Assuming a first-order dependence, estimate the overall rate of disinfection. Comment on the efficiency obtained compared to ultra-sonication in the previous problem.

29. Total coliforms in 50 litre of untreated water were found to be 8800 total coliforms/ml. This water is subjected to hydrodynamic cavitation at 3.4 and 5.1 bar. The count obtained at the end of 15 min was 5400 and 3200 total coliforms/ml, respectively. Estimate the overall disinfection rate at both the discharge pressures. Operating at which discharge pressure would be favourable?

30. Faecal streptococci in contaminated water was reduced to 7000 faecal streptococci/ml by using an orifice plate in a typical hydrodynamic cavitation setup at a discharge pressure of 5 bar. The count obtained under the same conditions without an orifice plate was 9980 faecal streptococci/ml. Estimate the overall disinfection rates in both cases, given the initial count of 12,000 faecal streptococci/ml. Comment on the result.

31. One hundred millilitres of contaminated water containing an initial microbial load of 6300 CFU/ml was subjected to ultrasonication using a probe type sonicator. After 15 min of treatment the microbial load was found to be 3200 CFU/ml. A similar experiment was conducted in a high-speed homogeniser and after 15 min of treatment the microbial load was found to be 1200 CFU/ml. If the rate of energy dissipation for the ultrasonic horn is 7 J/s and for the high-speed homogeniser is 45 J/s, which method is efficacious and why?

32. One hundred millilitres of contaminated water with an initial microbial load of 200 total coliforms/100 ml was subjected to ultrasonication using a probe type sonicator. After 15 min of treatment the microbial load was found to be 90 total coliforms/100 ml. What would be the cost per litre for reducing total coliforms completely (to zero) if the electric power consumption is 240 W and cost is Rs 3/- per kW.h?

Sample calculation:

Energy efficiency:

Time of treatment = 15 min (900 s)

Volume = 100 ml

Electrical consumption = 240 W

Initial microbial count = 200 total coliforms/100 ml

At the end of 15 min, microbial count = 90 total coliforms/100 ml

∴ Total coliforms killed in 15 min = 110 total coliforms/100 ml

Total coliforms killed/W of power consumed

= 110 CFU total coliforms killed/100 ml/240 W

= 0.458 total coliforms killed/100 ml/W

$\therefore$ Total coliforms killed/J of power consumed

= total coliforms killed/100 ml/W × volume/treatment time

= 0.458 total coliforms killed/100 ml/W × 100 ml/900 s

= 0.458 total coliforms killed/100 ml/J/s × 100 ml/900 s

= 5.09 × 10^{-4} total coliforms killed/J

Cost of treatment:

Energy efficiency = 5.09 × 10^{-4} total coliforms killed/J

To reduce total coliforms from 100/100 ml to 0/100 ml, energy required will be = 100/5.09 × 10^{-4} = 196463.65 J/100 ml

$$= 1{,}964{,}636.5 \text{ J/l}$$
$$= 1{,}964{,}636.5 \times 2.7778 \times 10^{-7} \text{ kW.h/l}$$

Considering 1 kW.h = Rs 3/– = 1,964,636.5 × 2.7778 × 10^{-7} × 3 Rs/l

$$= 1.6 \text{ Rs/l}$$

33. Two hundred millilitres of contaminated water with an initial microbial load of 500 faecal coliforms/100 ml was subjected to ultra-sonication using a bath type sonicator. After 15 min of treatment the microbial load was found to be 300 total coliforms/100 ml. What would be the cost per litre for reducing faecal coliforms completely (to zero) if the electric power consumption is 120 W and cost is Rs 3/- per kW.h?

34. Polluted surface water with an initial microbial load of 8000 total coliforms/100 ml was subjected to hydrodynamic cavitation. After 20 min of treatment the microbial load was found to be 1900 total coliforms/100 ml. What would be the cost per litre for reducing total coliforms completely (to zero) if the electric power consumption is 1950 W and the cost is Rs 5/- per kW.h? The volume of water treated is 50 l.

35. One thousand millilitres of contaminated water with an initial microbial load of 239 faecal streptococci/100 ml was subjected to ultrasonication using a probe type sonicator. After 15 min of treatment the microbial load was found to be 110 faecal streptococci/100 ml. What would be the cost per litre for reducing faecal streptococci to zero if the electric power consumption is 240 W and cost is Rs 4/– per kW.h?

36. Faecal streptococci in contaminated water were reduced to 7000 faecal streptococci/100 ml by using an orifice plate in a typical hydrodynamic cavitation setup at a discharge pressure of 5 bar. The count obtained under the same conditions without an orifice plate was 9980 faecal streptococci/ml. Estimate the overall disinfection rates

in both cases. The initial count is 12,000 faecal streptococci/100 ml. What would be the cost per litre for reducing faecal streptococci to zero if the electric power consumption is 2000 W and the cost is Rs 3/– per kW.h? Compare the cost of treatment and comment on the result.

References

Jacangel, J.G., Laõãneâ, J.-M., Cummings, E.W., Deutschmann, A., Mallevialle, J., and Wiesner, M.R. (1994). *Evaluation of ultrafiltration membrane pre-treatment and nano-filltration of surface waters* (90639). Denver, CO: AWWA Research Foundation.

Tchobanoglous, G.T. (1997). UV disinfection: An update. Presented at Sacramento Municipal Utilities District Electrotechnology Seminar Series, Sacramento, CA.

Tutunjian, R.S. (1985). Ultrafiltration processes in biotechnology, in comprehensive biotechnology, Vol. 2. In C.L. Cooney and A.E. Humphrey (eds.), *The principles of biotechnology: Engineering considerations*. Elmsford, NY: Pergamon Press, p. 417.

Viessman Jr., W., and Hammer, M.J. (1993). *Water supply and pollution control*, 5th ed. New York: Harper Collins College Publishers.

Index